WHO KILLED ALASKA?

WHO KILLED ALASKA?

Dave Brown

and

Paula Crane

NEW HORIZON PRESS
Far Hills, New Jersey

Library of Congress Catalog Card Number: 90-50876

Dave Brown
Paula Crane
Who Killed Alaska

ISBN: 0-88282-069-9
New Horizon Press

DEDICATION

I am dedicating this book to Betty: my wife, my lover, and my best friend. Together, we have shared almost three decades of thick and thin, usually too much thin and not enough thick.

I knew when I started this fight that it would get worse before it got better. I didn't realize how much worse.

If I had it to do all over again, knowing in advance exactly what I had to face, with nothing to ease the pain, I would gladly do it again if the same lady were still with me.

In spite of all the heartache, I still think the "man upstairs" has been pretty good to us, better perhaps than we deserve.

"In Wilderness is the preservation of the world."

—Henry David Thoreau

AUTHOR'S NOTE

This is the actual experience of a real person, Dave Brown. The personalities, events, actions and conversations portrayed within the story have been reconstructed from extensive interviews, diaries, and research, utilizing court documents, letters, personal papers, press accounts and the memories of participants. In an effort to safeguard the privacy of certain individuals, the authors have changed some names, time set and added a few elements and, in some cases, altered otherwise identifying characteristics. Events involving the characters happened: only minor details have been altered.

CONTENTS

PROLOGUE

Dave Brown brushed the stubborn, graying cowlick from his eyes and peered out the airplane window expectantly. All he saw were fluffy white clouds beneath the Boeing 727. "I'll be damned glad to get back to Alaska," he muttered to his neighbor. The man, heavy-set and balding, absentmindedly nodded and continued reading his newspaper.

But Brown wanted to talk. "The state ran me out of here three years ago, and now I'm coming back," he insisted. "I'm determined people are going to know what's happening to this beautiful place."

Not waiting for an answer, Brown rushed on. "When Betty and I were forced to leave Alaska, we carried everything we owned in our suitcases, in several foot lockers, three duffel bags, and two airline dog kennels. For the next six months, we slept on the floor again like we had done twenty-one years earlier.

"What we have been subjected to, so that the oil industry can maintain its veil of secrecy surrounding its illegal disposal and the destruction of Alaska's environment, has been rough. What we have been subjected to by the state of Alaska, so that they could cover up their own deliberate nonfeasance, and so that they could continue to appease the oil industry from which the state receives eighty five percent of its revenue, has been even rougher. But rough as it has been, it has not proven fatal. The oil industry and the State of Alaska both took their best shots. Now it's my turn.

"I'm not a two-week wonder, tourist journalist. This is my story and Alaska's. I have earned the right to tell it."

The neighbor raised his eyebrows, glanced sideways at Brown, but didn't respond. Brown noticed the man was reading an article about the ARCO-sponsored health fair in Alaska. Dave lapsed into silence, keeping an eye on the white blanket below.

Finally, there was a break in the white, and poking from between the misty particles was a jagged, granite-colored mountain peak, diamond-studded by the sun. Brown's heart flip-flopped. He knew the beauty that was soon to be spread before him—the familiar Chugach Mountains with their rainbow-lit glaciers; Turnagain Arm with its perilous mud flats that have enticed and, too often, claimed the lives of cheechakos (as natives called newcomers) unfamiliar with the ways of the wilds; and, of course, the booming development of Anchorage, spreading from pastel-shaded subdivisions perched precariously on the hillsides of the Chugach Mountains and on down to the moss-covered mud flats of Cook Inlet.

"Alaska's so damned beautiful," he commented to his neighbor. "It's a shame the oil companies are leaving it a polluted mass of garbage for our grandchildren," Brown sighed. His neighbor continued reading.

As the tips of mountains and vistas began to come into view, Brown recalled the first time he came to Alaska, twenty-four years before.

PART ONE

1

ANCHORAGE ARRIVAL

In March of 1966, Dave Brown, Betty, his wife, and children, Jeffrey, Jennifer, Mike, Deena and Jay, left Seattle and headed for Alaska. In their two ramshackle cars, they carried everything they owned: dishes, blankets, pots and pans, personal effects, sleeping bags, clothes, and Dave's hunting and fishing equipment. They had very little cash, so they brought lots of bread, baloney and other sandwich material. They slept in the cars, and they ate in the cars, because they couldn't afford motels and restaurants.

Most people drive the Al-Can Highway from Seattle to Anchorage in about three days. It took the Browns nine. Betty was terrified driving through mountains she had never seen before and on roads covered by ice and snow. At any rate, she kept up a steady fifteen miles per hour, and they didn't get any speeding tickets.

It was winter on the Al-Can. In the day, it was ice-cold and, at night, it got freezing. But with seven people, all huddled in one car to sleep, they managed to stay warm. They had one flat tire, luckily right in front of a gas station, so it was promptly fixed. At another point, while snow fell in a gray, bleak twilight, Betty's car slid off the road into a ditch. Fortunately, she slid to the west where there was a ditch and a high bank. Had she slid to the east which dropped into a canyon, they would have never made it to Alaska together. The car would have dropped out of sight.

Sliding off the road wasn't to be the Browns' only adventure on the highway. A few mornings later, after waking up from a frigid night huddled together in the old Studebaker, they were greeted by a developing snowstorm. White was everywhere, but they decided they had to get going, because there was no shelter where they were and at least the edge of the road was still visible. This time Dave was in the lead, inching along. By afternoon, the snow was falling harder and Dave kept looking for a road sign indicating food or shelter ahead. For hours he searched, but there were no visible turnoffs. His eyes began watering from the strain. Finally, off to the left was a set of faint wheel ruts in the snow. Dave followed them with Betty right behind. They ended up in a small field where, dead tired, they parked for the night.

During the night, the snow stopped falling, and the sky cleared, but the temperature dipped to twenty degrees below zero. The next morning neither vehicle would start.

The kids were clamoring for breakfast, so Betty began feeding them while Dave walked around in the bone-chilling morning wondering what he should do next. He scanned the horizon looking for any kind of sign of life, and saw a column of smoke up the road about a half mile.

"I think luck is with us," he said to Betty. "There's some kind of building with people in it up there. Stay here with the kids and I'll walk up and see if I can get any help."

Trudging up the road, little bits of snow stinging his face, Dave found a rustic log highway lodge. Inside, the caretaker and his wife, a rough-hewn couple, he big and crackly, she like a caricature of Mrs. Santa Claus, were warming the place up to get it ready for the tourist season.

"Welcome," the man said, as he spied Brown. "You must have some sort of problem."

Dave explained his problem, along with the fact that his wife and five children were back in his cars.

The caretaker's wife was sympathetic. "My, my," she said. "You shouldn't be staying out there in the cold. If you had been able to drive just this much further last night, we would have let you stay here in the lodge for free. We haven't seen a soul in three weeks!"

The caretaker loaned Dave a propane tank with a torch and a stove pipe. Dave walked back to the vehicles, thinned the oil in the oil pans, and both vehicles started right up again and they were soon back on the road to Anchorage.

The snowy, flinty road, bordered with boulders and small avalanches, was hazardous, but the soaring, snowy peaks, the canyons, glistening lakes and streams in the distance thrilled Dave in a way he had never felt before, making his heart surge. He looked around at his family and felt they'd all been blessed with a God-given chance to be worthy of that heart-pounding beauty.

They had no more mishaps. On a sun-sprinkled day, Dave proudly drove down Fifth Avenue into the burgeoning city of Anchorage. Stopping in a Fifth Avenue parking lot, they bought the kids each a soft drink and then Betty and Dave walked through the back door of a Fourth Avenue bar. There they bought an Anchorage newspaper, a cup of coffee for Betty, and a bottle of Oly (Olympia beer) for Dave.

The prices for the paper and coffee weren't bad, but the bottle of Oly was a dollar and a half. The bartender was a big, husky woman about sixty, and she looked tough enough to whip a grizzly bear with her bare hands. "Look," Dave complained, "a buck and a half is a hell of a price for the bottle of Oly. In the supermarkets in Seattle, I can buy a half a case for a dollar and thirty-five cents."

She answered Dave gruffly, "Sonny boy, this ain't Seattle, and this ain't no supermarket."

6

ANCHORAGE ARRIVAL

The cost of beer was just the first shock to jolt Dave and his family. When Dave and Betty went back out to the cars and the kids with their newspaper and checked the classified for a place to rent, they discovered that rents were incredible, at least for them. In Seattle, they had rented a very nice, four-bedroom house for ninety dollars a month. In Anchorage, there was nothing for less than about four hundred a month. They didn't have enough cash left for anything like that.

Finally, Dave found a tiny advertisement from a man named Marvin Byer, who owned an old trailer court off Tudor Road on Piper Street. He wanted someone to manage his trailer court, collect his space rents, and do some necessary maintenance work. In exchange for doing the work, he would rent them the manager's house for a mere sixty bucks a month. It took Dave three days to convince Marvin Byer that he should have the job and the house. Beyond the low rent, the job paid nothing. During those three days, Dave and his family slept in their cars in the snow banks on the edge of town. On one day, they stopped for coffee at the old Milky Way Diner on Fourth Avenue. While they were inside, someone broke into one of their cars and stole the blankets. Luckily they had a few more in the other car.

The reason Marvin Byer took so long in deciding Dave's family's fate was because his last trailer court manager had scampered down the Al-Can with all of Marvin's space rent. Dave and his family had just come up the Al-Can, and after nine days on the highway—plus three days sleeping in the car on the edge of town—they looked pretty rough.

Despite his misgivings, finally Marvin gave in. "Here's the sixty dollars rent," Dave said slowly, realizing how low his cash was running.

"You owe me eighty dollars more for a utility deposit," Marvin said.

Dave swallowed hard and paid him the eighty dollars. Then Dave headed for the nearest pawn shop with his deer rifle so that his family would have money for their first groceries.

WHO KILLED ALASKA?

When Betty and Dave took their first look at their new home, they had a further shock. It was a two bedroom, flat-roof, do-it-yourself house built by someone who was not very good at doing it himself. "It's not much more than a shack," Dave said to her sadly. "I wish I could do better for you and the kids." There was no place to sit in the house except on the toilet or at the kitchen table, which only had two chairs. So they sat on the floor and they also slept on the floor. It was, Dave admitted later, rough, but they survived. Alaska was where they wanted to be and they had each other.

A few weeks after moving in, Betty let out a happy squeal and pointed out the kitchen window. One of their neighbors had just bought a new couch and chair and the people had put their old couch and chair out by the side of the road for the trash truck to pick up. It was trash collection day. Dave and Betty grabbed that old couch and chair barely ahead of the truck.

Someone must have seen them because shortly after that, a squat, salt-and-pepper haired lady knocked on their door. "I hope you're not offended, but I have two sets of G.I. steel bunk beds. I'd like to give you them for the kids," she said.

"We're not offended at all. We're grateful," Dave said humbly. He and Betty took the bunk beds and, shortly thereafter, they acquired a used bed and dresser for themselves.

Within that week, Dave got a job as a new car salesman for Alaska Sales and Service. They put their kids into school and Betty went to work as a cashier in a grocery store in the neighborhood. While working at Alaska Sales and Service, Dave struck up a friendship with Bill Mears, another salesman. A few months later, Bill left Alaska Sales and Service and went to work as a general manager for Construction and Drilling Equipment Company. He came and asked Dave if Dave would like to go to work for him selling heavy equipment. Dave immediately agreed.

Because Dave and Betty believed it would happen, and

because they worked hard making it happen, their situation gradually improved.

"Those days were tough," he murmured. "But they were the happiest days of our lives."

2

CHEECHAKOS

While the Brown family encountered some amazing prices and spartan conditions in their shift of life to the far north, Dave also found exactly what he was looking for—a pristine wilderness with superb hunting and fishing.

It was during these early times that Dave and friend, (and subsequent employer) Bill Mears, headed down to the Russian River on the Kenai Peninsula. It was shortly past noon on a brilliant, cool spring day that Dave and Bill hiked down the power line to the river, and headed toward the big, clear, aquamarine pool. Dave's arms started twitching in anticipation. This was why he'd come to Alaska.

The Russian River was the ultimate in salmon fishing for freezer-owning anglers. The river is a tributary of the glacier-silted Kenai River, and as the sockeye salmon head up the river, suddenly emerging from the gray-green silt as they travel toward their spawning ground, anglers can see them and target them, and better still, the salmon can see the flies. The Russian River is open to fly fishing only, so anyone fishing with any type of rod and reel just has to put a fly on the end of it.

10

CHEECHAKOS

Dave and Bill were true fly fishermen. They came armed only with their fly rods and brightly colored buck tail flies, known as "coho flies."

Since sockeye salmon travel upriver in schools, when the water is clear, it's easy to know when anglers really need to be on their toes. Bill was fishing down river from Dave, when he suddenly whooped, "Here they come!" That meant Dave had to cast his line out where they were, and be ready to drift it just right—under the fish's nose. Salmon have only one thing on their minds when they are heading upriver—sex. So a fisherman has to play on the emotions of the fish, not on its stomach. A sockeye will not bite at a fly out of hunger, but because it happens to be there as he's passing through, mouth gaping in true fish fashion. Because placement of the fly is so important, having clear water is also important.

Dave watched nervously as the first shadows of fish began heading his way. He heard another whoop from Bill. "Fish on!" Bill shouted. Dave glanced down river. He could hear the fly line on Bill's rod screech out of the reel, but the fish was headed back down river. Dave sighed in relief. He was far enough away from Bill that he wouldn't have to take his rod out of the water, just when the fish were passing through.

Dave's first drift through the school came up empty. He tossed the fly back upriver and watched again as the bright yellow buck tail floated within inches of a sockeye's mouth. He couldn't let Bill have the best of him, even though he knew Bill was a top-notch fisherman.

Bill was still fighting his own fish when Dave tossed the fly upriver once again. Suddenly he felt a tug. His yell of pleasure had to be louder than Bill's. "Fish on," he yelled, copying Bill. He hoped his fish was bigger. He cupped his hand around the reel, giving the fish just enough line to fight. A glinting silver form jumped out of the water, and Dave estimated it to be about a ten-pound male. Steadily, Dave began to carefully reel in his fish. Fly reels can be quite tricky. One has to maintain just the right drag to keep the fish on, without giving him so much line he runs with it, or too little line that he snaps it. Dave kept his tip up and his line tight when suddenly the fish made a break toward him. "Hey, cut it out!" Dave squealed and reeled in the surplus line frantically, but it was too late. The sockeye leaped in the air, spit the hook and seemed to smirk at Dave on the way down. Dave glanced over at Bill and noticed sheepishly Bill was landing his fish, still too preoccupied to notice that Dave had lost his. Trying to be nonchalant, Dave examined his line, to make sure that everything was all right, then tossed it back into the water. He tried to be philosophical. The school was still coming, and losing a fish was just part of the game.

Luck proved him right. Dave got another bite almost immediately, but felt that something was different about this fish. "This must be a really tiny female, or maybe even a jack," Dave thought to himself. A jack is an immature salmon who, for no known reason, decides it needs to swim upriver with the rest of the mature fish. They're usually under sixteen inches.

About a minute or two later, Dave landed the fish and realized the source of his embarrassment. He had landed a Dolly Varden. He was going to quickly release the fish back into the water, but the small char had swallowed the hook, and to release it would be to release it to die. It would be better to take it home and eat it, even if the small catch would embarrass him in front of Bill. Anyway, Dave rationalized, Dolly meat is pretty darned good, and Dollies had once even had a bounty on their heads because of their propensity for eating salmon eggs.

Bill walked up, laughing. "That's quite a monster you caught there," he said, dangling his ten-pound salmon conspicuously.

"Don't worry," Dave replied. "I'll get my limit yet."

Bill scratched his head and smiled. "Oh, I know you will. The river is loaded with fish right now. I'm heading back to my spot to cash in."

Bill was right. The school of sockeye hadn't quit. They kept on coming, out of the silty waters and into the clear. Nervous fear and pleasurable anticipation nuzzled at Dave as he tossed his fly, once again, into their midst. He simply couldn't go home with nothing but a Dolly, when there were so many salmon passing by.

The next salmon was his. Dave yelped with pleasure. Within two hours, both men had landed their limit, and were relaxing on the shores of the river before driving the two hours back to Anchorage. It was one of Dave's first fishing trips in Alaska, and his first introduction to the Kenai Peninsula, and the area tantalized him. That spring, Dave was again thrilled at the flats at the mouth of the Kenai River. The air and ground were filled with thousands of snow geese making their annual rest stop before going on to their breeding grounds in the far north.

Dave took a boat trip through the inland passage, checked out the sand dunes in the Katmai National Monument and observed the bears at McNeil River. At summer's end he borrowed a canoe and paddled down the Swanson River and the Moose River canoe trails, seeing beautiful lakes, moose, loons, and trumpeter swans.

Dave bought a riverboat and called up a friend, Malcolm Stickly, nicknamed "Stick," and invited him to join Dave in his riverboat to garner the silver, glinting salmon which should have by then been swarming into Twenty-Mile River.

Dave, Stick and Stick's son launched the boat and headed up Twenty-Mile River, a silty-gray river which drains the broad, flat Twenty-Mile glacier. They came to a crystalline, running creek, and Dave said, "This is it. I have a feeling the silvers are here."

They headed up the creek, which in places was almost as narrow as the boat. A mile or so up the creek, they came to a long, deep, clear pool. Dave could see flashes of silver in the shadows along the edge of the left bank. He cut the motor and eased the anchor down quietly. The three men began to rig their lines, and Dave's was in the water first. Instantly, he hooked and landed a twelve-pound, gleaming fresh silver.

"Now this is fishing," Dave exclaimed as he cast again. At the end of each cast, there was another fish. Every fish ranged in weight from ten to twelve pounds. They soon had their limit.

Stick spotted a canoe with two moose hunters in it, and the three quickly tried to cover their catch. As the canoe drifted by, the men asked if they had had any luck.

"Nah . . . there's nothing here," Dave lied, hoping they wouldn't look into the depths of the pool and see the silvers that were swarming there.

Fish that throng so thickly in such small streams can be quickly decimated, both by greedy anglers and by thoughtless ones roaring up the creeks, cutting up fish with their motors and eroding the riverbank and surrounding habitat. After the canoeists left, Stick made a wide, sweeping gesture, "This is so fragile." That comment stuck in Dave's mind.

By the spring of 1967, Dave had made enough fishing and hunting buddies to truly begin exploring Alaska. One of the prerequisites for in-depth Alaskan exploration is an airplane, and Dave was thrilled that Paul Yeoman, one of his new-found friends had one. Often, the three of them—Paul, Ray Docharty, who later became his business partner, and Dave—would travel the skies and the trails in search of game and excitement. One day, they were flying over the Chistochina River, counting moose. In a forty-five minute flight, they counted more than fifty moose, half of them bulls. It would be a perfect place to hunt during hunting season. They spotted two of the bulls fighting on the west bank of the river. They buzzed them with the plane, and the two bulls broke their fight long enough to run across the river and resume their fight on the other bank.

Armed with the knowledge that the moose were there, Dave and Betty, and another couple, Dick Sjoden and his wife Barbara, decided to hunt up the Chistochina River. They headed upriver in Dave's track vehicle, an Army weasel, vintage 1946. Late in the evening, the weasel broke down, leaving the two couples stranded with a burned-out generator and voltage regulator. They were twenty-five miles up river. In the morning, Dick and Barbara announced, "We're hiking out." Betty and Dave nodded in understanding, "We'll stay and try to repair the weasel and then ride out." Unfortunately, it didn't work. They finally gave up and decided they'd walk out the next day.

As they walked for hours, stumbling and falling, along the Alaskan Muskeg, Dave turned to Betty, "What the hell is this? It looks solid, but, in reality, it's pocked with sink holes." Betty looked miserable as she first stepped on a tiny knoll that looked like it would support her weight, then sank up to her knees in ooze. "My new boots are ruined," she cried. Soon, her feet began to be more of a problem than the muskeg. But she limped on in misery, her feet blistered and bleeding, knowing there was no other way out of the predicament.

After hiking another twenty miles, they heard an airplane. Dave grabbed Betty's arm in support. "Maybe we'll get a ride somehow," he said, trying to instill courage into his wife. Betty tried to smile, but the pain was too excruciating.

Soon, the Super Cub buzzed overhead and the pilot, an absolute stranger, hung his head out the window and yelled, "Meet me at the river."

Feeling more hopeful, Betty limped with greater speed through the brush and timber to the river. There, she and Dave met Jim Cann, a bush pilot, guide and concerned Alaskan who risked his neck landing on a gravel bar in the middle of the river just to help out Betty and Dave, two strangers.

Super Cubs are two-seated airplanes known for their adept ability to land and take off on the back of an Alaskan mosquito. Even so, the river gravel bar was brush covered, so Jim took out his ax and cut the brush on one side just to make sure his take-off, and re-landing later to pick up Dave, was a little safer. Then he flew Betty to Chistochina Lodge, carried her into the building and helped her take off her boots and bloody socks.

Once reunited at Chistochina Lodge, Dave and Betty couldn't thank Cann enough, but Cann refused any payment for his work. "I'm a bush pilot," he insisted. "I know I'm the only way out in a situation such as you were facing, and that's why I do it." Knowing Alaska was full of such good-hearted people began to affect Dave in a way he'd never felt before, even more than hunting and fishing. His ties and feelings of allegiance to his newly found home grew and multiplied. Even his problems cemented this bond.

Hunting in Alaska often gets hunters into trouble, and Dave's friends were no exception. On one trip when Dave couldn't go, his friends Ray and Paul shot a moose near the Chistochina River, then borrowed an airplane from another friend, Joe Bear, and Joe's nephew to recover it. While taking off from a gravel airstrip near a miner's cabin—more than fifty miles from the nearest highway—Paul cracked up the plane, catching a wing tip on the brush after being blinded by the bright sun during takeoff.

Paul and Joe's nephew were cut up and bruised a bit, but they walked on foot about five miles and came to a cabin and airstrip belonging to guide Eldon Brandt from Eureka. Eldon was getting ready to fly out and he gave them a ride.

Ray and Paul came to Dave, asking if he would help them salvage the plane. The plan was to take Dave's weasel, dismantle the plane, and use the weasel to bring it out.

The plan was simple, the kind cheechakos often make. The only problem was that, like most newcomers, Dave didn't know that making plans which involved the wilderness can be complicated by nature, both its lure and reality.

When Dave left home, he told Betty he'd be gone about four days. Then Dave and his two friends headed across the wilderness toward the crashed airplane.

When the three men arrived, they quickly surveyed the scene. The plane had crashed next to an unoccupied, but comfortable, old gold miner's cabin. There were two ancient, weather-beaten, horse-drawn sleighs behind the cabin, and plenty of old, weather-beaten lumber. The three men took the front runners and tongue from one sleigh and built a wooden bed to go in it, similar to a pick-up truck bed. Then they put two cross beams on the sled, projecting out about a foot on each side of the bed, one at the front of the sled, and one at the rear. Ray was the engineer of the job. He could build something out of almost nothing.

Unfortunately, the first complication was Dave himself. Dave got sick with diarrhea, and he only left his sleeping bag for one reason. But, after a couple of days, he finally felt good enough to grab his rifle and head across the creek and up into the hills after he had seen a caribou.

While the job was supposed to be dismantling an airplane, three hunters can't be in the middle of a large supply of game, in the middle of hunting season, without hunting; besides, Dave rationalized, he had five hungry mouths to feed at home, and Ray had eight.

The caribou disappeared, but he heard some ptarmigan clucking, then saw a cow moose and a calf. Dave watched as the cow and calf slowly walked by within fifty feet. Then, with his thoughts turning to roasted ptarmigan, raising his rifle, he shot a couple. Almost immediately afterward, the sound of hooves beating the ground caused Dave to whirl around. Mama moose didn't like the sound of those shots, and was charging toward the threat to her calf's very existence.

"Hey! Hey! Slow down!" Dave yelled at the mother moose, hoping to distract it. Then he raised his rifle, and had her in the sights. "Come on, mama," he said calmly. His shout had lowered her speed to a walk, so he continued talking. "I'm not going to hurt your little baby. And I don't want to hurt you, either, but then again, I think just a little bit more highly about my hide than yours, so if I were you, I'd remove your hide from these cross hairs." The moose stopped within fifteen feet, the hair on her back standing straight up, then turned, and, with her calf, slowly walked away. That night the ptarmigan had an extra flavor of excitement in them.

By now, the four days had stretched close to two weeks, but the golden leaves of autumn, the ice cream chill of frosty days and the thrill of hunting kept beckoning to them. The dismantling of the plane, as is usual in Alaska, took much longer than the three men had envisioned. A few mornings later, as they looked across the creek, they saw a bull moose and a grizzly bear on the hillside. Nearby, a herd of migrating caribou grazed in the valley below. There were no thoughts of heading for home. Instead, the men pulled on their boots, grabbed their rifles, and tumbled out the cabin door, chasing game. The moose and bear disappeared quickly, but the caribou remained. They shot one cow caribou to eat and six big bull caribou to take home to their families.

As the three men were dressing out their caribou, the Civil Air Patrol flew low over them, and then returned to Anchorage to inform the news media, and the wives, that the wayward men were not lost or injured, merely alive and having a good time.

And a good time it was. Toward twilight, they finished up the butchering and sat down for a few sodas. Suddenly, a big bull caribou, with a huge set of antlers, walked toward Ray, snorting. Clowning, Ray picked up a set of antlers from one of the dead caribou and put them on his head. Both female and male caribou have antlers—male antlers are just quite a bit larger. Dave watched, first amused, then concerned, as the bull bent down and pawed the ground and snorted. It didn't matter if the bull thought Ray was a male or female. He could still do damage. Ray kept playing with his antlers until the bull was within twenty feet of him, snorting. Finally, Dave shouted, "Ray, if that bull wants to fight you or make love to you, he can have you. I'm not going to kill him to save you."

Ray looked sheepishly at Dave and dropped his antlers, shooing the bull away.

At dawn the next morning, the men hooked their sled to the weasel, loaded the meat into the sled, covered it with a tarp, laid the airplane fuselage lengthwise on top of the meat, put the two wings on each side of the sled and lashed the entire load together with rope. They put the airplane engine inside the back of the weasel, put the antlers on the roof, and called it a day.

Then they started up the mountain. The ground was bare of snow, but the arctic chill had moved in. With the heavy load, it took the entire day to ascend the peak and another day to go back down the other side; so they stopped to spend the night there. The cold was even more bitter the next day, and their problems were just beginning. Within fifteen or twenty minutes of starting, the transmission was torn out of the weasel. The load was just too much for it. The three men took their rifles, their sleeping bags, filled several pieces of visqueen with cooked caribou meat, and headed out. By then it was 7:30 in the morning.

The temperature continued to plummet all day. Soon there was ice on every creek. The men had to bash the ice in with their rifle butts to get some of that cold, clear water. Then the ice grew so thick they had to shoot holes in it. At nine o'clock that pitch dark night, shining their flashlight beams in front of them, they were still walking, when they passed a beaver pond. The dam was about five feet high, and Dave could hear water running—good drinking water they could have without having to blast a hole through the ice.

"I've been a hunter, fisherman and a back country wanderer ever since I was a boy," Dave said smiling. "I have always been a thirsty wanderer. Whenever I cross a creek, I stop and drink." He bent down and scooped up some of the water with his cupped hands.

"No telling how far it may be to the next water hole."

Ray and Paul stayed and rested on the trail, while Dave walked the fifty feet to the beaver dam.

After drinking his fill, he yelled, "Come on, guys. You better get some of this water while you can."

"Shut up," came Ray's reply.

Dave was a bit flabbergasted to be talked to like this. "What do you mean, 'Shut up,' " he called back.

"Shut up and look," Ray replied, pointing to the sky behind them.

Out of the sky came a bright, phosphorescent, glowing, dome-shaped object. It was round with a flat bottom and dome-shaped on top. It came down quietly out of the sky in kind of a lazy zigzag fashion, following the trail. It stopped and hovered at tree-top height, not more than one hundred yards behind them, and directly over the trail.

"Holy shit," Paul murmured. "What the hell is that? A UFO?"

Dave said with a half serious, half silly smirk, "If it is, they probably wonder what kind of idiots would be out for a walk so late on such a cold night in such a desolated spot." Despite Dave's attempts at humor, each had an uneasy feeling.

Dave eased his nerves by attempting to make conversation. "Shall we walk up closer for a better look? Should we shoot it down with all three of us shooting at once, or should we run and get the hell out of here?"

They all stood riveted to the spot and stared.

After a few minutes, the thing streaked up into the heavens and out of sight, traveling faster than a jet.

(Once back home, Ray reported the incident to his commanding officer in the Air Force at the time, and found out there were several reported sightings of UFO's in Alaska at that time.)

They couldn't think of or find any rational explanation for what they'd seen, so the incident did little to settle the nerves of three men who had been hiking in the cold woods for fourteen hours. "That's it for the day," Ray called out. The other two nodded. They pulled out their cold caribou to settle the growlings in their stomachs and began to gnaw at it. Then they all crawled into their sleeping bags. Despite the excitement, all three were so tired they fell asleep almost immediately.

By 7:30 a.m., they were on the trail again, gobbling caribou, blueberries and cranberries which were growing profusely along the trail. But the cold and the poor diet were beginning to take their toll, and their hiking speed was severely diminished.

Around noon, a plane flew over and landed nearby. The pilot was a friend of Ray's, who had heard the three men were still not home, and had come out to check on them. He brought a big box of raisins, and the men charged into it like they'd never seen raisins before. The snack gave them the extra energy they needed to finish their hike that afternoon. Once again, Dave was impressed with the Alaskan spirit of helping others.

During the winter that followed, the weasel did make it back to civilization, but it wasn't easy. After installing a new transmission, Ray discovered that with the snow on the ground the strain on the weasel was not so great, but the traction was not as good, either. He made it as far as the river, but the river was too deep and not frozen over, so he couldn't cross it. He was forced to leave the weasel and the sled, with the load, at the river bank.

The three men went back in that May and recovered the weasel and the gear, but all six caribou were gone, and the evidence of wolverine and fox hair left no doubt as to what stole it. Dave sighed. At least the meat wasn't wasted. He was sure a caribou smorgasbord was held on the riverbank by the wild creatures that got there first.

As 1968 drew to a close, Dave was self-employed in the plumbing and excavating business and he made friends with Frenchy Lamoreaux, one of Alaska's oldest and, in Dave's opinion, finest guides. In early spring, when the frost is not yet out of the ground and Dave's excavating business couldn't quite get into full swing, Dave went to work as an assistant brown bear guide at Frenchy's camp on Kodiak Island. To Dave, it was a vacation with pay.

Most of the hunters were gritty and enthusiastic and a pleasure to guide. Dave learned to behave like a native, eating fresh dungeness crabs and digging clams from the sand when the tide was out, sometimes getting five or six little butter clams in a single shovel. He mastered getting the dungeness crabs in shallow water like the others did, wading in his hip boots.

But Dave was most fascinated by the bears. Frenchy's main camp was on Kaiugnak Bay, with a smaller tent camp on Selua Bay. They spent a lot of time sitting down on a ridge or a hillside where there was good, long-range visibility, scanning the other hillsides with binoculars, looking for bears. Dave once counted fifty brown bears during one two- or three-week hunt, although they were probably the same bears seen over and over again. Dave would sit and watch through the binoculars, fascinated by seeing the bears playing like children. He watched them sit on their rumps in the snow on a high hill and slide to the bottom like a child on a flying saucer. He watched as they would climb back up to the top and do it again.

Dave found nothing fascinating, however, about carrying a wet brown bear hide on his back in a pack board while walking in hip boots, with his rifle slung over his pack board and sweat running down his spine until it found its way into his long johns and ran down his butt. That was just plain hard work.

While actually catching the fish and shooting the game is a nice touch for an Alaskan adventure, it is really the adventure which endears the state to the heart of a true outdoorsman, and Dave had become enthralled.

During the spring and summers of the next years, Dave floated the Matanuska and the Chickaloon River in an Avon raft. He paddled through both the Swanson River Canoe Route and through the lakes along the way. Later, he ran the Kenai River in his river boat and in his canoe. He ran his river boat the full length of Skilak Lake and Tustemena Lake as well as Lake Louise, Susitna Lake, and Tyone Lake. He crossed the Cook Inlet many times in that same river boat, from Anchorage to the Little Susitna River and to the Big Susitna River, usually around early September with his 30.06, his shotgun, his spinning rod and his labrador. Dave never cared whether he shot a moose, some ducks, or whether he caught a few silver salmon.

With Dave, it was always the journey that mattered rather than the destination. There was no climax to any trip. He loved every minute of every journey he ever took. There was always something interesting or humorous around the next bend. And humor is a necessity, because sometimes the trips don't turn out as planned. However, they become good tales to tell in later years, as long as one lives through the adventure to be able to tell the tale. As Dave grew more comfortable with the land he heard and saw that some people do not.

CHEECHAKOS

Tustemena Lake is a lake on the Kenai Peninsula—a twenty-six mile long, innocent-looking lake—which invites unwary boaters to explore the headwaters. The lake is fed by the huge glaciers of the Harding Icefield, which has a reputation for hurling its winds across its ice and over the normally placid waters of the lake, whipping them into a frenzy. Dave was doubtful the lake could truly live up to its reputation. He had spent the first thirty-one years of his life on the shores of Lake Ontario, and had served a long stretch on a destroyer in the United States Navy, so he scoffed at the tales Alaskans told about how rough their little puddles could get.

One day Dave, his son George and John Wood, his friend and attorney, headed up the lake in Dave's twenty-foot riverboat. The water was calm, and they set up camp at the head of the lake, pulling the bow of the boat up onto the sand and tying it securely tied to a big log. After hunting for a few days, they saw little sign of moose but they did notice a tent camp up the lake shore with a bear hide nailed to a tree trunk, so they had some hope.

During the night, the bristling wind began to sing, and Dave grabbed his flashlight to check out the boat. First glance showed him the boat had vanished. And a closer look revealed it had sunk, still securely tied to the big log. The stern was completely under water, and the waves broke over the top of the motor. The steering console was broken loose from the floor boards.

"John! George! Help me salvage the boat!"

The two other men groggily followed him, running down to the shore. "Hey, why'd you get us up?" John protested. "There's nothing anyone can do except salvage the loose gear floating around. You could have done that yourself," George snapped. Looking more closely at it, they all agreed and went back to bed.

The lake had calmed down by morning and the three of them began bailing and, with a lot of grunting and groaning, got the water out of the boat. "I don't think the motor will start with all the water in it," John said. "I'll hightail it to the camp down at the lake shore," Dave offered. "Let's hope the people are still there."

They were, but they were getting ready to leave. Their boat was lighter than Dave's, so they had pulled it high above the water, avoiding Dave's problem. They carried Dave back to the parking lot, then he drove back to get a second boat and headed back down the lake toward George and John. He met them halfway. They had dried out the plugs, got all the water out of the cylinders, and had the old Johnson humming.

That wasn't the only time Dave's boat sank. That November, Dave, Howard Keiser and Dennis Safranek went on a deer hunting trip on Whale Island, which lies in the Pacific Ocean between Kodiak and Afognak Island. The trip got off to an ominous start. They loaded the four-wheel-drive pickup with their gear, attached the twenty-three-foot inboard boat, and headed down the black-ice covered road toward Homer.

Black ice driving is a skill Alaskans need to learn if they are to survive. It requires extremely steady nerves which transmit themselves down to an extremely steady foot. Any change in speed or braking will send a vehicle into a spin. Dave was heading down a long, straight hill when five moose stepped slowly out onto the right side of the road and began to amble across. By the time he reached them, two of the moose had stepped off the road on the left, and three were still standing, blocking the left lane, when Dave careened by. He could have reached out his window and patted them on the rumps. It was a scary beginning of a scary trip.

When the men launched the boat off Kodiak Island, they could see Whale Island in the distance. It was late in the afternoon and getting dark, along with the beginning of a snowstorm, was veiling the island. Howard was the guide. He had lived on Kodiak Island for several years, and had often hunted on Whale Island. Even so, Dave was leery about traveling the open ocean to the island with darkness stealing in.

Howard reassured him. "Don't worry. I know exactly where we are going and we will be there before you know it."

By the time they reached the island, darkness and a full scale snowstorm nearly obliterated the land. Howard instructed them. "We are looking for a bay with a high buff above it. There's an old deserted cabin and we'll stay there."

They headed up the coastline, and soon Howard yelled, "This is it. Pull in here. This is the bay we want, and there's a good gravel beach you can pull right up on."

Dave cut back on the throttle, and headed slowly, blindly, toward shore and ran the boat up on a boulder the size of a house. It was the wrong bay, the wrong bluff, and there was no cabin. The boat was solidly hung on the rock, and they were facing the wind, the waves, the snow, and the dark. On top of that, the boat had sprung a leak.

Howard and Dennis leaped into the frigid, chest-deep water to push the boat off the rock. Dave pushed with a paddle. When the boat broke free, Dave reversed the engine, then got out of there and headed full speed for a light he'd seen in the distance. Howard and Dennis waded to shore and built a roaring fire in a dead spruce tree.

By the time Dave made it to the lighted cabin, his boat was half full of water. Knocking on the cabin door, he explained his predicament to Paul Singer, the commercial fisherman who owned the cabin.

"We'd better get Tom. He's down the road," the weather-beaten, blond-haired fisherman pointed thoughtfully.

Tom, Paul's friend, took a skiff and retrieved Howard and Dennis. Then, Paul lit the light on board his own boat, building a fire in the boat's stove.

"Look, you're welcome to stay here as long as necessary," Paul offered. They agreed and he took them to his house for hot coffee and a hot meal, so that Howard and Dennis could dry out.

The three men stayed several days in Paul's boat, repairing the damage to their own boat. In the meantime, their hosts went deer hunting and came back proudly with two large bucks. Dave, Howard and Dennis never got off a shot.

Still, Dave loved his Alaskan adventures. They were exciting and, sometimes, a bit aggravating, but always added up to a good tale to tell later on, without conjuring up fearful memories.

WHO KILLED ALASKA?

But that isn't always true for Alaskan adventures. The state has the highest accidental death rate in the nation. A lot of people do thoughtless things which put their own lives in the hands of a sometimes angry Mother Nature. The Brown family wasn't immune to that bad judgment.

Dave put a lot of faith in his wife's decisions, and often his own predicaments were the result of failing to listen to Betty. But even wives can be wrong sometimes.

One summer, Dave took Betty, the two youngest children, Jay and Deanna, his Labrador retriever Oly, two rifles, a shotgun, fishing tackle, his wall tent, groceries and gear into his riverboat, and headed across Cook Inlet from Anchorage to the Little Susitna River. Moose season opened the first day of September, and Dave wanted to be ready for moose, ducks, or silver salmon. They motored up the Little Su River, and made camp at a bend in the river. Jay and Dave hunted moose for a couple of days while Betty, Deanna and Oly relaxed around camp in the gorgeous weather.

Dave had plans to stay a week or ten days, but somehow they lingered and while they were at the camp, the first day of school came and went and Deanna started getting really anxious to get back and see her friends. She began to bug her mother and Dave constantly to go back home, and became—like all teenagers—slightly insane in her methods of getting attention. Once, she climbed a tree, yelling down, "I hope I fall out and get hurt so that I have to go back." When that didn't work, she threatened to beat on her arm with a stick, hoping to break her arm. Of course, her brother interpreted, "only if she could do it painlessly."

Finally, Betty couldn't take it anymore. "She's driving me crazy," Betty informed Dave. "Let's go home—now."

Dave looked across the sky. It was filled with black, speedy clouds. "Betty, up here we're sheltered from the wind, and the weather seems just fine. But take a look at those clouds. Something's moving them. I don't think it would be safe to try crossing Cook Inlet right now. We'd better wait until the weather calms down."

"Aw, come on," Betty chided Dave. "You've never let a

few clouds keep you from doing anything before. You just want to stay, and I'm sick and tired of listening to Deanna. You're away from her during the day, hunting. I have to sit here and listen to her, all the time."

Finally, Dave gave in. They broke camp, loaded the boat and headed down river. As soon as they hit the mouth of the river, he swung the boat toward Anchorage, and safe haven in Ship Creek.

But once they got to Cook Inlet, Dave knew they were in serious trouble. Six to ten foot waves were coming up the inlet, along with an outgoing tide—Cook Inlet boasts the second highest tides in the world—high winds and darkness. Dave looked at Betty, who, by this time, was rigid with fear. "Turn back," she pleaded. "I wish to hell we could," Dave said. He, too, wished he had never come but turning against the waves would mean swamping the boat. The huge waves had begun breaking over the stern corner of the craft. Dave tried to crank up his 75-hp motor.

"I can't do it all the way or I'll swamp," he said, shivering. "And I can't run it slowly either, or the tide will carry us down the inlet—away from everything."

For a long time, the little boat was idle in the water, all the while filling up with water. Dave watched helplessly. "I have to somehow get that bilge drain unplugged, but I can't let go of the wheel."

"Jay," he yelled. The fifteen-year-old boy looked up at him, fright washing over his face. "Somehow, you're going to have to get to the back of the boat and pull the drain plug. If you don't, we're going to swamp."

Like Dave, Jay saw the gravity of the situation. But trying to move around in a riverboat tossed nearly into the air by waves is damned difficult. Jay crouched, grabbing at everything in sight, and inched his way toward the stern.

"I'm almost there," he yelled, reaching into the water. "I've got the plug." Everyone breathed sighs of relief as the water began to drain. Then Dave saw ominous shadows in the night and suddenly realized they were ships, waiting for the storm to subside before docking in Anchorage. Riverboats don't have lights, and can easily be squashed by a huge tanker.

"Jay," Dave yelled again. "There's a flashlight in that duffle bag. Get it out and shine it at those tankers. Circle it around and around. Maybe they'll be able to avoid us."

Dave knew, however, that the main responsibility for avoiding a crash was his. Tankers have trouble turning and stopping quickly. He muttered, "Lord, please help us out of this one and I'll never bother you again."

The run between Little Su and Anchorage usually takes forty-five minutes. Four miserable, cold and terrifying hours later, the little family pulled safely into Ship Creek and the downtown Anchorage boat harbor.

During these early—and best—years of Dave's life in Alaska, Anchorage was a small city—about 34,000 people were clustered in various subdivisions in what is known as the "bowl," formed by the Chugach Mountains on the east and Cook Inlet on the west. The Brown family had finally settled in a comfortable ranch house in a part of the area known as Rabbit Creek, about ten miles from downtown Anchorage. Their neighbors were few and scattered throughout the wooded hills of the Alaskan subdivision—which consisted of a network of dirt roads and electricity connecting a few houses. Their life, if somewhat secluded, was cozy and pleasant. Frequently a moose would wander down their driveway, sometimes even a brown bear sashayed through the area, eating dog food.

The moose weren't always a welcome sight. Children in Alaska are constantly reminded that moose are dangerous animals that can stomp you to death if they so desire. So there were many occasions when Dave would have to escort his children to the school bus stop because of the moose which had decided the Brown's front yard was the place to rest. The Rabbit Creek subdivision was built in the moose's natural wintering area, and the moose had no concern that they may instill fear in the smaller two-legged creatures. And they didn't like the four-legged creatures that barked and seemed to stay near the two-legged creatures. In fact, Dave once saw a cow moose and calf chase the neighbor's dog up onto their porch, then stand at the bottom step and keep the terrified animal cowering on the porch. No one was home to let the poor dog in.

While Dave's comfort in Anchorage increased, so did his comfort with the surrounding countryside, so that he now felt quite capable of taking other people out to share in the enjoyment of his wilderness. The atmosphere was seeping into his psyche. He was becoming like the "true Alaskan" he admired.

One time he offered to take Eddy, the teenaged son of his wife's boss, on a black bear hunt. Dave and Eddy flew into Cache Lake in the Talkeetna Mountains, which offered a tremendous view of Mt. McKinley. They set up camp in the ruins of one of Frenchy's old cabins—a cabin that had seen one too many snow loads on its roof. The hunt wasn't to yield any meat for the table, or hide for a rug, but it did yield the wilderness experience which is more important than the meat and hide.

The next afternoon, after a hard day's hunt, Dave and Eddy chopped their way out of a thick patch of willow and into an open meadow. Once there, they stuffed themselves with enormous blueberries and bittersweet cranberries.

"Why don't you go over the top of the hill and see if there are any bears over there," Dave instructed Eddy, then he settled into the grass for a quick snooze. An hour later they both hiked back to the camp.

The next morning, they climbed to the top of the high ridge behind the cabin, and started scanning the surroundings, looking for black bears. Both pairs of binoculars soon focused in on the meadow where Dave had taken his nap the day before. Suddenly a beautiful bull caribou ran out of the brush and across the meadow. He was running for his life.

"Keep your eye trained on the spot where the caribou came out, and, in a second, you'll see a bear."

Suddenly he yelled, "There's a huge grizzly running at full speed after the caribou."

Eddy called out, his voice shaking, "Where?"

Dave said, trying to keep his own voice calm, "Over there. In front of the brush."

By then, the bear realized he had lost the race with the caribou, stood on his hind legs, and actually shook his right forepaw at the caribou. It was a sight any teenager would carry throughout his life.

In 1970, Dave's oldest son, George, graduated from high school in upstate New York and decided that he wanted to come to Alaska to spend two or three weeks with his father. He was then supposed to return to New York to enroll in the local community college. When George came, he brought his old Springfield 30.06 and his fishing tackle. When Dave saw George, Dave laughed, "It must run in the blood."

Dave took George, Jeff and Mike back to Cache Lake, and this time the hunt was successful. George shot a black bear, and Dave shot a bull moose. Mike and Jeff caught some rainbow trout and, amazingly, George caught a big rainbow with a chunk of moose meat. The experience was enough to convince George that Alaska suited him. He stayed, and is still living in the state he, like his father, loves.

Both Dave and George found a way of life—a quality of life—in Alaska that, as they put it—"they could never have found anywhere else in the United States."

During these years, Dave changed. He had come to Alaska with his family just as the pioneers moved west—for a better life. But, as time went on, Dave grew to love the land. Thriving on the hunting and fishing, enthralled by the fantastic scenery and wildlife viewing, and responsive to the helpful and generous pioneer spirit around him, Dave became an ardent defender of anything Alaskan.

One of Dave's pet peeves was people's misconceptions about "The Great Land." When he would go "Outside" for vacation, people would tease him about living in a land of ice and snow, and wonder how he could keep his igloo warm.

"Why in the world would you want to live in Alaska?" a

friend would ask. "What kind of place do you live in? What in the world do you do with yourself all winter when it's dark all the time? What a boring place to live! Sure, there are mountains, but so much of it is just wasteland, from the pictures I've seen."

Dave was all too ready to answer, patiently, but indignantly. "I do not live in an igloo. My house is just as nice as yours. I heat with fuel oil, and I have two fireplaces and a wood stove just for the pleasure of enjoying wood heat once in a while. I have indoor plumbing. In fact, almost everybody in Alaska has indoor plumbing."

Dave would then reflect on life outside of the city, in the Bush, where mostly natives or pioneer-spirited white people live. "Sure, there are some who live in cabins and heat with wood. Their lifestyle is quite rustic, but it's by choice, not by chance. They came to Alaska to get away from the complex life of an industrial nation. The natives that live in the Bush even live in regular homes, although there are more outhouses in the Bush.

"I have never seen an igloo, and I don't know of any other Alaskans, including Eskimos, who have ever seen an igloo, although I hear they are still occasionally used for temporary shelter on long hunting trips, that is, if the Eskimos don't have REI tents. The only igloos I ever saw were the ones I built as a kid in grade school in the Lower Forty-eight, together with the neighborhood kids. We built them from a picture we had seen in the *Weekly Reader.*

"And you think there is nothing to do! Anchorage, Fairbanks and Juneau are as modern and up-to-date as New York, Phoenix, Los Angeles, Chicago or Seattle. We have concerts, movies, dances, art shows and many restaurants. There's a McDonald's on almost every corner in Anchorage."

"But it's got to be cold! I know it's cold!" the friend would repeat.

"Sure, it's cold," Dave would reply simply. "It's cold in Minnesota, too. In fact, many times, while watching the weather, I would notice that the temperature in Anchorage would be much higher than the "low" in the Lower Forty-eight. In fact, I've seen wintertime temperatures in Anchorage that were warmer than places in the south, like Arkansas or Texas.

"I'm leading you on a bit," Dave would admit. "It does get bitter cold in the Interior and Northern Alaska. But Anchorage and the Kenai Peninsula's weather is tempered by the ocean. It's never really as cold as in the northern Midwest or the Northeastern states. We get a lot of snow, but not as much as the Rocky Mountain states, although there are places that it really comes down, like Valdez.

"That's all winter weather. In the summer, the Interior can get up to the 90's and a person can get just as sunburned as in the Lower Forty-eight."

As Dave became more and more "an Alaskan," he began to tout the wonders of glorious life in Alaska and to tell his fair share of tales about the cold wilds of Alaska.

"While working as a cat skinner at forty or fifty degrees below zero, I've had to drop my pants when nature called and had all of my fingers curl up like claws from the cold by the time I got buttoned up again. It's a real effort at fifty degrees below to peel off your parka and your insulated pants, drop them, then drop your levis, and last of all your long-john drawers. By the time you reverse the entire procedure, your fingers are pretty well curled up, and you can't wait to get back on the seat of that old D-8 and get moving again. Your nose runs, your eyes run, and even though it's fifty degrees below, the glare of the sun off the snow is so bad that you wear sun glasses. You can't stop your eyes from running, and the tears freeze on your sunglasses starting at the bottom and build toward the top. Finally, you can't see out of them any more, so you take them off and put them under your parka and down vest next to your shirt so that your body heat will warm them up, and you can wipe them clean and start all over again. You ask yourself, "What am I doing here when I could be home with my wife in front of a warm fire with the coffee pot on the stove and her bringing me free refills."

While the land is beautiful, there is still some treachery. Dave's oldest son, George, while operating a D-7 cat, broke through the ice on the Colville River on Alaska's Frozen North Slope. He went to the bottom in fifteen or twenty feet of water. The ice, or the water pressure, broke the glass out of the front of the cab, so he climbed out the front window onto the hood, then up on the cab roof. Then he made it to the edge of the ice and climbed out. He broke a tooth or two and wrenched his back, but he was back at work again in a few days, same place, doing the same thing. George was lucky. Every once in a while, they lose a good man that way. Sometimes, when they go through the ice, it's in fifty to one hundred feet of water instead of fifteen or twenty.

Along with the stories of the cold and treachery of wilderness, Dave likes to talk about the beauty of winter, too, specifically the shimmering, colored northern lights. The lights dance and prance across a clear, cold night, and the experience of being treated to such a show is indescribable.

Dave's fiery defense of Alaska had begun to extend to episodes on his job, too, where he often met people who were in the state to get hefty paychecks, shipping everything to some state Outside, where they would eventually ship themselves when the job was over. Dave had little patience with those type of people whom he believed were parasites, sucking on Alaska's vibrancy.

One day, while working on the Trans-Alaska Pipeline as a relief operator at Livengood Camp, his foreman, a man from Alabama, drove him to his worksite.

"Look at all that worthless bottom land. What is this country good for, anyway?"

Dave huffed indignantly, "At this very moment there are more than twenty thousand people working on that pipeline, most of them sending home the biggest paychecks they had ever seen in their lives! If Alabama is so great, why are you in Alaska, anyway?"

Of course, Dave's staunch defense of the state did not earn him points with his bosses or companies, most of whom were not from Alaska.

When Dave was at Terminal Camp in Valdez, he was working as a 977 cat loader operator for Fluor Alaska. He ate lunch every day in the dry shack with seventeen operators. Fourteen of the seventeen were from outside the state of Alaska, the result of a "brother-in-law, cousin and good-old-boy" hiring network. There's always a troublemaker in every job, and the crane operator from Seattle was this job's complainer. He complained about everything about Alaska—the worthless wasteland, the rain, the snow, the school tax, Alaska unemployment tax—everything. Finally, Dave could listen to no more.

"If you don't like Alaska, leave it. We don't need your kind up here. It's a good thing you didn't bring your family up here, because one of your kind is certainly enough. There are planes leaving Valdez every day. Why don't you get on one of them?"

Then Brown turned to the others in the dry shack. "There are thirteen other non-Alaskans here. If any of you guys thinks like he does, then why don't you go with him?"

It wasn't the first time Dave's sharp tongue got him in trouble, and it wouldn't be the last. The company trumped up some other reason to fire him (they couldn't fire him for speaking his mind) and handed him a pink slip.

When Dave Brown believes he's right, he'll go to all lengths—in fact, he'll just about do or say anything—to prove it. His problem with Fluor was just a taste of what was to happen between Dave and the state of Alaska. One could say it gave him some practice for the big time.

The pink slip stated that Brown was "inept" because he had broken a pipe inside the compressor building with his 977 Cat loader. He checked out the pipe. It was still intact, and Dave's tracks showed he hadn't gone within three feet of it.

Once back in Anchorage, he filed suit against Fluor through the Alaska Department of Labor. But he needed evidence, so he sneaked back into Terminal Camp on three separate occasions, staying two or three days each time. Friends hid him in their rooms and brought food. During these illegitimate sorties, he gathered thirty-four handwritten statements. Each statement said that Brown was a good and careful operator, that he was helpful and cooperative at all times, and that he had not broken the pipe he was accused of breaking. Further, Dave contended angrily, if he was so inept, why did Fluor keep him on the payroll for almost a full year?

Charged with energy, facing a foe whom he believed was harming his state, Dave welcomed the resulting hearing in Valdez. Fluor had come armed with several witnesses who were still on the Fluor payroll. However, Dave had previously gotten statements from them, so after their pro-Fluor testimony, Dave's attorney whipped out these statements, read them, then asked each to verify the statement.

Though it cost four thousand dollars in attorney's fees, Dave delighted in the result of the hearing. The hearing officer had decided in Dave's favor and against Fluor. He flatly stated that Dave had been discriminated against because he was an Alaskan, and because he was proud enough to speak out in defense of his state. Dave Brown became the first white Alaskan to win a discrimination suit against a pipeline contractor.

PART TWO

3

OIL—BEFORE AND AFTER

Dave Brown never lived in Alaska "before oil." But he did live in Alaska before the oil boom which resulted from construction of the pipeline from the North Slope to Valdez. By that time, the Kenai Peninsula had already been given a good dose of then-legal drilling muds into the earth. Anchorage experienced the full pipeline boom in the mid-1970's. After Dave moved there, the Kenai Peninsula had its own growth spurt.

But the peninsula did not experience the full force of oil development, with the natural accompaniment of a mushrooming state bureaucracy and its resulting jobs and sprawling development, until about 1983. So Dave was able to live in both Anchorage, and then the Kenai Peninsula, before they exploded with people and problems.

Until this era, Alaska was getting along fine with minimal oil development. Its economy was based on the United States military, commercial fishing, logging, mining, construction, guiding, trapping, and a growing tourism business.

OIL—BEFORE AND AFTER

Many of the people who came from other states and settled in Alaska were pioneer types—individualistic and adventurous. Anyone who wanted to work could usually find work. These early settlers, unlike the drifters who came to Alaska strictly for employment, came to Alaska to live because they wanted to become Alaskans, not because they wanted to get rich. They liked the outdoor lifestyle and the opportunity that the land offered them. It was a serene atmosphere, made to order for folks who didn't care for life in the fast lane. A man wasn't judged, or given a job, on the basis of his family tree, nor by his position in the social register, nor by where he'd been or what he'd done. All that mattered was a person's character. If your word was good, you were good. You were accepted. A lot of people like Dave Brown, who had problems or didn't fit in in the Lower Forty-eight, came up to Alaska. The state and its people gave them fresh starts and the opportunity to succeed when they had failed elsewhere.

Oil changed everything.

Oil was first discovered in the state at Katalla, not far from Cordova and the Gulf of Alaska, in the 1920's. A refinery was built at Katalla, but it burned down in 1933, and was never rebuilt.

In 1923, a well was drilled at Cold Bay on the Alaska Peninsula. It was a dry hole. Other locations were tested, and, finally, in July 1957, oil was discovered on the Kenai Peninsula at Swanson River by the Richfield Oil Company. Richfield's well was a gusher, and from that moment on, Alaska would never be the same.

After Swanson River's beginning, construction of offshore drilling platforms in the Cook Inlet began at a feverish pace. Then, by the late 1960's, two major industrial companies began to take shape on the Kenai Peninsula. The Collier Carbon and Chemical Company started building a chemical plant near Kenai to produce ammonia and urea for fertilizer, using the excess reserves of natural gas that existed on the Kenai

Peninsula. The plant went through several name changes, and today it is known as Unocal Chemicals Division. Shortly after Colliers, Phillips built its liquefied natural gas plant. The gas is compressed at extremely low temperatures, transforming it into a liquid state. It is then shipped across the ocean to Japan in refrigerated tanker ships. Two other refineries were then built—the Tesoro Refinery, which refines crude oil into gasoline and jet fuel, and the Chevron Refinery, which mostly makes asphalt, along with other oil products.

Today there are fifteen offshore drilling platforms anchored to the bottom of Cook Inlet by steel legs, serving four hundred and sixty-five wells that have been drilled in the bottom of the inlet. Oil and natural gas flow from these platforms through pipelines to refineries and other onshore installations. Some goes to Kenai, some goes to Anchorage, and some of it is loaded onto tanker ships for shipment to the Lower Forty-eight and to Japan and Korea. There are also currently three major on-shore oil and gas fields—Marathon's Beaver Creek field which is about five miles from Kenai's city limits; Chevron and ARCO's Swanson River oil field which began the first oil rush; and Unocal's Kenai Gas Fields just outside the city limits of Kenai and the Cannery Loop unit, in the city limits. Across Cook Inlet there are several other oil support facilities—Drift River, West Forelands, McArthur River, Nicolai Creek, Granite Point, Beluga River and Ivan River. The Drift River terminal, with its storage tanks, was built at the base of an active volcano, Mt. Redoubt, and recently had to undergo major flood prevention work in order for the state to allow it to continue to operate. A 1990 eruption caused some flooding of the storage tanks, and it was feared a major eruption could cause a major oil spill into Cook Inlet.

During the late 1960's, while Dave was selling heavy equipment to the construction and oil industries, he got a taste of what it takes to do business with the oil industry. The company for which Dave worked was the Alaskan distributor for a particular brand of welding machines. The company sold a lot of those machines to a major international oil field contractor. Although there were only two salesmen working for the company, and Brown was one of the two, Todd, the lanky, dark-haired general manager of the company, said in a secretive tone, "Dave, don't go near the head man of that oil field contractor. The account with that contractor is one which I handle personally."

Dave and Todd became pretty good friends, fishing together, and tilting quite a few bottles of Oly together. One night, after more than a few Olys, Todd became more talkative, "Dave, you probably have guessed there are some kickbacks involved in doing business with that particular company." Todd hiccuped as he went on, "Those welding machines sell for a certain price to anyone who wants them—the manufacturer's suggested retail price plus the freight to Anchorage. But, for you-know-who, we sell the welding machines for the listed price plus a little extra." Dave was a little shocked, but he leaned forward to listen more closely. "Is there more I should know?" Dave asked.

"Probably not," Todd said, "but tonight I'm gonna" he slurred the next word, "tell you."

"The bills go directly to the contractor's home office. When they're paid, I pay the head man with the contractor his take." Dave's company benefited only in that it was able to sell the welding machines to the contractor. The contracting company itself paid a slightly higher price than it should. The head man with the contractor was the one who pocketed the extra cash.

"Why would you go along with that kind of nonsense?" Dave asked, puzzled. "You don't make enough to take the risk." Todd confided to Dave that although he really didn't like doing it, if he didn't, the contractor would just buy another brand of welders from another source instead.

Not too long after this discreet conversation, Dave and Betty were shopping in J.C. Penney's, and saw Bart Jamison, the top man with the oil field contractor buying new furniture and appliances for his new home.

As Betty and Dave watched speechless, Jamison rolled back a wad of one hundred dollar bills and paid for the furniture and appliances in cash. Dave made a mental note of the transaction, and couldn't help but wonder how many of those bills were kickback bills.

Dave continued to learn bits and pieces that made up the interesting, bureaucratic and corrupt nature of the oil business. He noticed that when he sold a major piece of equipment to a non-oil contractor, the question always before the contractor was how was he going to pay for it. However, whenever Dave sold a major piece of equipment to an oil field contractor or an oil company itself, paying for it was never even a consideration. Their question was, "How soon can we get it?"

Another major international oil field contractor had an office in Anchorage. They had a young purchasing agent named Ted Tanner, whose wife was still in the Lower Forty-eight. He didn't know anyone in Alaska, and he was lonely. Dave's wife was an excellent cook, so Dave invited Ted over for dinner. At the same time, Dave had a neighbor who was out of town working at a remote location. Janet, his young wife, was lonely and frequently came over to visit Betty. On the afternoon of the day Ted came to dinner, Janet was there, and she eventually invited herself to stay for dinner. Dinner turned into a long evening after which Ted took Janet home. The Browns later learned he spent the night. After that night Dave could do no wrong with Ted's company, but Dave didn't like it. Ted was never asked for dinner again, and Janet was never allowed to hang around Betty. Dave wanted no part of an irate husband who might just come home and choose to do away with the perpetrator or accomplices in his wife's adultery.

With oil came men, without wives, and women lonely for their husbands, so one-night stands became commonplace in Alaska, and slowly morality crept down the gutter. Topless and bottomless bars appeared overnight, along with plentiful massage parlors.

While the Swanson River and Cook Inlet platforms brought the first oil boom, the real big one—Prudhoe Bay—was still to come. One Alaskan, Tennessee Miller, hitched himself to the oil boom and, as owner of Frontier Rock and Sand, pioneered for the oil companies all over the North Slope with cat trains, pulling men, equipment and supplies wherever the oil companies wanted them to go. The pioneering paid off for Tennessee and it paid off for the oil industry. On April 22, 1967, Atlantic Richfield had spudded in a well named Prudhoe Bay State #1. Fourteen months later and 12,016 feet deeper, on July 24, 1968, ARCO and Humble Oil and Refining Company publicly announced that a major oil discovery had been made at Prudhoe Bay on Alaska's frozen North Slope. Both companies had brought in two spectacular wells near Prudhoe Bay, just seven miles apart from each other. Humble Oil and Refining Company later became known as the Exxon Company.

4

JOB PROFILE

Dave worked on the Trans-Alaska Pipeline for almost three years as a heavy equipment operator. He terms it "the most interesting fiasco I have ever been allowed to participate in."

The waste in time, money, material and labor was so ridiculous that a normal person couldn't even begin to fathom it. The problem was, according to Dave, "the way it was bid." If it had been a "hard money" project, each phase of the job would be put up for bid with engineer estimates and the jobs would go to the lowest bidders provided they were qualified—that they had the manpower, the equipment, the capability and the bonding to get the job done. In those kinds of jobs, the contractor who is awarded the job hurries to get it done as efficiently as possible, so he can get paid and go on to another job.

But the pipeline was built as a "cost plus" job, where the philosophy is to put twice or three times as many men, or twice or three times as much equipment, on the job as are needed, and then go as slow as possible, because the longer the contractor takes, the more money he makes.

The contractors were all paid "cost plus" by the Alyeska consortium of oil companies which built the pipeline, and Alyeska—and therefore the oil companies—were paid "cost plus" at the gas pumps by the American people. Dave reported that most of his fellow workers had the feeling they weren't working the pipeline to get it done, but just to take up time and money. Dave saw criminal waste of equipment—materials, supplies and labor at all the camps he worked—Valdez, Livengood, Five Mile Camp and Coldfoot.

At Valdez, he walked out on the job one crisp morning with a fellow worker, Don Peters.

"Look over there," Don said, pointing.

Dave looked in the direction. Two crane operators were fighting a duel along the dock with their cranes, tipping one of them over and bending it out of shape.

"What the hell do you think they are doing?" Dave protested.

"It's all in the name of fun," Don retorted sarcastically.

Dave's face wore a grim look, "Well, I guess we and the rest of the American people will pay for that stupidity at the gas pumps."

At Valdez, there were two 977 caterpillar loaders owned by the contractor. Dave operated one, and a man called Goose Loring operated the other. Goose would sit all day in his loader with one pane of glass broken out of the front of his cab. Occasionally, he would lean out through the missing window and drop an empty soft drink can down his exhaust stack. Then he would climb back in the operator's seat, pull back on his throttle, and see how high up in the sky he could blow the empty can.

Later, Goose was put on a bobcat loader. His favorite trick was to drive the loader inside an empty container van, closing the doors behind him. There he would hide all afternoon. Several times during the winter, his foreman followed his loader tracks through the snow to the doors of the van, and kicked him out. But nobody ever fired him.

There was really no point to firing him. All he was supposed to do was to stay awake, anyway. There was no real work for him to do.

Dave did his best to stay busy. If he couldn't find any dirt to move, he would carry tools for the laborers, steel for the iron workers or oxygen and acetylene for the welders. Dave felt he earned his pay, but it was hard. He had to look for work.

One day, on his way to the main gate, he passed a small pond with a 225 cat backhoe next to it. The operator and oiler, at two hundred dollars an hour, were playing with the machine, killing time. With the backhoe, they had put a big rock in the little pond. On top of the big rock they had placed a medium-sized rock, and on top of that they were trying to balance a small rock, sort of like building a snowman with rocks. It was a silly game, but, considering the concept of cost plus, it wasn't senseless. The contractor was getting paid only for the hours that machine was running. It didn't matter what the machine was actually doing.

Dave was older than a lot of the men on the job, and brought to the pipeline some of his 1950's views, including anti-drug opinions. Although he drank his share of beer, his hard drug views got him a quick "red neck" label. But whereas Dave's pay day night beer never got in the way of his performance, the men who smoked pot and snorted cocaine on the job could, with their drug-induced incompetence, endanger everyones' lives, including Dave's. Once, he watched a pot-stoned surveyor in Valdez step back too far while using his transit, and fall into a twenty-foot deep hole. He survived, unhurt, but a lot of the men on the pipeline weren't so lucky. Carelessness and overconfidence, much of it under a drug cloud or while hung over, caused many deaths.

One of Dave's friends, Jess Straight, was killed at Valdez. In his case, it was his own overconfidence, not drugs nor booze. A heavy equipment and diesel mechanic—and a good one—he jacked up a truck and crawled under it without setting any blocks. The jack slipped and Jess died.

A laborer lost his arm at Valdez his first night on the job because the crane he was working with was operated by the oiler and not by the operator. Alyeska normally held safety meetings once a week. The workers could always tell when someone was killed, because there would be an extra safety meeting. Dave and his buddies would first hear the ambulance screaming through camp, and then there would be a safety meeting—like a death knell everyone tried to, but couldn't avoid.

At Livengood Camp, Dave worked as a relief operator, covering other operators while they were on R & R (rest and relaxation) or while they were sick. As a relief operator, he operated 'dozers, scrapers, loaders and, for a while, a six-inch Gorman Rupp pump. Operating this pump opened Dave's eyes to the absurdity of a "cost plus" job. The six-inch pump was used to fill a shiny yellow Kenworth water truck, and it was situated on the bank of Hess Creek. For two weeks, while the pump operator was on R & R, Dave ran his pump.

A six-inch Gorman Rupp pump has an electric starter. This pump also had a suction hose down in the creek, and a discharge hose mounted high in the air on a pipe frame. The water truck backed down under the discharge hose. All Dave had to do was turn the key on the pump, then push the starter button and wait five or ten minutes while the tank filled with water. When it was finished, he would turn off the switch. He did this three or four times every day. But when he wasn't turning the key or pushing the button, he was sitting down in the dry shack reading magazines. He worked a total of not more than half an hour three or four times a day, yet got paid for twelve hours a day, seven days a week, and cleared slightly more than $1,000 a week, after taxes.

It wasn't just Dave. It took four men every day to get that water truck filled. A teamster drove the truck because that was a teamster's job. The teamster couldn't run the pump because that was an operator's job. Dave was the operator. The

Kenworth water truck had two bucket seats; a laborer rode on the other seat. The laborer had to climb on the side of the truck's water tank and open the lid. Later, he had to climb back up and close the lid. That's three men. In addition, the operator, Dave, had to get to and from the dry shack from Hess Creek. The foreman—the one who hated Alaska—would drive Dave three or four times a day to the creek, just so he could turn his key and push his button.

Union labor required the division of labor, and that added to the cost; that was the way Alyeska wanted it. Using union labor, the company could generate more cash flow and justify raising the price of gas at the pumps. When the job was completed, and the "cost plus" was gone, Alyeska got rid of the union labor and replaced them with less expensive, non-union workers, mostly from the Lower Forty-eight. That policy has continued until today, with the marriage between the unions and the oil industry nearly completely broken.

Life on the pipeline wasn't all irony and absurdity. There were just plain jovial incidents, and some simple hilarity. Dave always got a kick out of relating this story:

"At Valdez, a laboring crew was drilling rock one day, all of them covered with dirt and wearing coveralls, goggles, and hard hats. One of the men had to urinate and apparently didn't have time or didn't feel like walking to one of those little green plastic outhouses that Alyeska had scattered all over the site, so he unzipped his coveralls and started to go. One of his fellow laborers tapped him on the arm and said, 'Hey, do you see that guy over there in the coveralls? Well, under those coveralls, hard hat and goggles is a girl, not a guy!' The urinating laborer wet his pants trying to turn it off in mid-stream and put it away."

Women were hired on the pipeline, working as laborers, teamsters, operators, oilers, welders and helpers. Some were ornamental, some were foreman's and superintendent's helpers. Most were there to make money, and they worked hard.

At Valdez, the women were separated from the men. One two-story barracks was filled with women only. Another two-story barracks had women on one floor and men on the other. All the other barracks were filled with men.

The other pipeline camps in which Dave worked were smaller, so the arrangements were more haphazard. Dave's first night at one of the smaller camps gave him quite a surprise. At ten p.m. he made a trip to the men's room. There were four or five urinals in a row followed by a row of sinks along the same wall. No one was in the room that he could see, so he stepped up to the urinal next to the sinks. Just as he started to urinate, a female voice said, "Hi there. How are you?" A girl in her twenties stood at the sink, just three or four feet away, getting ready to brush her teeth. Dave zipped his wet fly and tried to look casual as he sauntered out of the room.

While women in camp sometimes created an invasion of privacy, there was a much more threatening invasion. Bears, both black and brown, were everywhere—in dry shacks, in dumpsters, in pick-up trucks and busses. They would even occasionally get into the mess halls and barracks. Dave watched once as a man dropped some trash into a steel dumpster in Valdez, not knowing there was a bear inside until the bear stood up. The man shouted and scrammed. A couple of black bears walked into a dry shack at Valdez while the men inside were eating lunch. That created quite a scramble.

Dave had his own bear problems at Livengood Camp. He was operating a D-7 winch cat near a steep hill. A mechanic had driven him to the foot of the hill where his dozer was supposed to be, but the night shift had instead left it on top of the hill. In the morning rain, the mechanic drove the truck as far as he could, but the hill was much too steep for a flat-bed truck to climb. They both noticed a black bear ambling about two hundred feet away.

"Are you afraid of bears?" the mechanic asked Dave.

"Heck, no!" Dave huffed proudly. "Bears and I get along just fine."

"Well, good. I'll just drop you off here and get back to

work. I can't climb that hill, and the foreman will be here shortly to drive you up in his four-wheel-drive pickup."

As the mechanic drove off, Dave instantly regretted his macho reply about bears. Here he was with no trees to climb, and no weapon except a stainless steel vacuum thermos. Sure enough, the bear decided to check him out, or at least try to see what was in Dave's lunch. But Dave didn't feel like sharing his lunch. Another quick glance at the local scenery revealed only a JLG manlift. Dave wasn't very familiar with a manlift, but he knew it required two people—one person on the platform, and another actually on the machine doing the lifting. The box platform had pipes on all four corners and safety chains all around and was about four feet off the ground—slightly safer than standing in the middle of an open field.

Dave climbed the platform, with the bear close behind, placing front paws on the edge of the platform and looking at Dave through his little beady eyes.

Dave shrugged nervously, "Coffee, tea or me, huh?" The bear was four feet away.

A quick glance around the platform revealed a case of insect spray in one corner.

"Hey, big fella," Dave said calmly. "There isn't room for both of us up here."

The bear waved his head back and forth. Brown bears may have poor eyesight, but Dave knew this one could smell lunch and, at four feet, he could see pretty well, too.

Slowly, Dave reached for a can of insect spray, then quickly hurled it just above the bear's head. The bear whirled around, chasing the can and biting into it. Bug spray is pretty nasty stuff; it sure made the bear mad. But it kept him away from Dave. So Dave threw another one, and another one. The bear chased after them like a dog. Finally, Dave reached back and realized he had hold of the last can. He was running out of luck. Then he noticed something else on the platform—a sledge hammer. He would make his last stand. Either he would kill that bear with a blow from the sledge hammer or die trying.

Just as Dave was reaching for the last can with one hand and for the sledge hammer with the other, his foreman drove up. He slowly rolled down the window, looked at all the bitten up cans and laughed till tears came to his eyes.

But Dave's sense of humor had been seriously imperiled by the ordeal, and he snapped, "Just shut up, run that bear out of here, and come pick me up!"

The foreman did run the bear off, at least a little ways. Far enough to be able to pick Dave up and take him to the top of the hill where two dozers were parked—Dave's and another operator's, Ron Caine. Two other pick-ups were parked with men waiting for the busses and drill rigs to come so they could go to work. Ron was in one of the pick-ups, but he had left his lunch in his dozer.

Pretty soon, that disagreeable and hungry bear, full of the taste of bug dope in his mouth, came ambling up the hill. He climbed up onto the D6 dozer, sat down in the seat, and promptly started to eat Ron's lunch. Run jumped out of his truck and ran to his dozer, thinking he could scare the bear off. But bears around there—accustomed to people—didn't scare very easily. This one just got out of the seat and climbed down the track, ready to defend his lunch from this harassing man. The bear won. After he had successfully chased Ron back to his pick-up, he put his front paws on the hood and glared. He must have recognized Dave as the perpetrator of the bug-dope melee, because the bear came over to Dave's side, jumped up, and pressed his nose against the glass of the window, this time just six inches from Dave. Dave grinned back, thinking he had won.

But the ordeal wasn't over. The Fourth of July was coming up and the company had planned a big celebration with fireworks. The fireworks were in the other pick-up. The men started lighting them and throwing them at the bear. The first three or four firecrackers didn't impress the bear much, so the men threw a cherry bomb directly in front of him. That got his attention, and he slowly walked away, then stopped, stood up and pawed the air, as if to say, "Okay, I'm going, but don't push your luck or I'll come right back."

WHO KILLED ALASKA?

The pipeline brought all sorts of people to Alaska; so it wasn't really unusual that Dave would meet an old friend and former co-worker from a Seattle job he had before moving to Alaska. Sten Berg was working as a mechanic on the same job Dave was. One day, at lunch time, the lunch bus parked in the gravel pit and all the men piled into it to eat lunch. Sten was in his mechanics truck, so Dave joined him there.

Soon, as if hearing a dinner bell, a huge black bear ambled out of the woods. Although it was against the rules to feed black bears, it was done regularly, and, as a result, the bears became dangerously accustomed to men, associating men with food—and not just a few measly scraps. Bears ate side orders of prime rib, steak, king crab, and, of course, sandwiches, so they soon forgot all about their supposed main course of berries, bugs and mice.

As this particular black bear sauntered up, he got what he wanted. The men in the bus dropped food from the side windows. But lunch does end, and the bear had not gotten its fill. It simply waited by the door. Soon the day began to get hot, so he crawled under the bus to get into the shade. Nobody dared to leave. A water truck arrived and sprayed water all around and underneath the bus. The bear appeared to enjoy the mid-day bath and he kept the men trapped in the bus for about an hour.

Dave and Sten were watching the shenanigans from the truck, also leery about stepping out. Finally, Dave turned to Sten and said, "Drive me over to my loader and I'll get rid of the bear before the whole crew gets fired."

Dave drove the loader to the side of the bus and shoved his four-yard loader bucket under the edge of the bus. He rocked the loader bucket up and down and shook the entire bus, almost like an earthquake. The bear wasted no time in getting out. Dave chased the bear a half mile across the gravel pit, but the bear didn't give up easily. He thought he might try sparring with that yellow monster. Three times, he whirled around on his hind legs, ready to charge the yellow machine.

"Damn fool!" Dave yelled, hitting the brakes to keep from running the bear over.

The pipeline construction induced one big feeding frenzy for the bears. It did the same for the people—people who fed on the huge amounts of money paid for relatively little work. It was a partying frenzy, boozing frenzy, and drug frenzy too. Yet, in spite of the unbelievable wages and incomparable rich banquet of food set before them, men quit their jobs every day. They didn't need an excuse. They knew that in order to get rehired, all they had to do was drop into their union hall.

Dave overheard a conversation one night through the paper-thin barracks wall that seemed to be typical of the mental attitude of the men on the pipeline.

"Hey, Joe. If we work seven days a week, we get to go on R & R in nine weeks. If we only work six days a week, we don't get any R & R until we've been here thirteen weeks. We're only working six days a week right now. Joe, have you ever worked thirteen weeks on the same job for anyone?"

"Hell, no, man. I ain't never worked thirteen weeks in my whole life and I sure as hell am not going to start now."

They probably quit after a while, went home to Texas or Oklahoma for a couple of weeks, then got rehired for a couple more weeks. A lot of the men earned more money on the pipeline than they could have earned in an entire lifetime. Many of these same men spent it on drugs, booze and good times. When the pipeline was done, they had nothing to show for their work in Alaska.

While the "cost plus" nature of the pipeline caused extremely high construction costs, other factors also contributed. One company, Ketchbaw, in charge of x-raying the pipeline welds, decided if one good weld was x-rayed several times, with different identification numbers, it might simplify their task. In the wake of this, Alyeska fined Ketchbaw and filed suit against the company. Ketchbaw counter-sued Alyeska. A man in charge of the investigation suddenly was found dead and the death was ruled a suicide, after which x-rays of three hundred and fifty eight welds were reported stolen from a pipeline camp office.

The x-ray incident was not resolved, but Dave was sure if Alyeska's supervision had been anything but inadequate, this incident and many other similar wastes never would have occurred.

Some of the high construction costs were blamed on pilfering of the equipment and tools by the workers. Dave witnessed a lot of stealing, but the majority of the thieves he saw were not underlings.

The working men on the pipeline project could not drive their personal vehicles inside the camp gates at any of the camps. Most of the workers left their cars at home in Anchorage or Fairbanks . . . or Texas or Oklahoma . . . and went by Alyeska bus to the pipeline camps. In order to get through the gates, they were subjected to a security guard's scrutiny.

Company officials, superintendents and foremen all drove freely through the gates in company pick-up trucks, Suburbans and Blazers. A large share of the stolen tools and equipment went out the gates in these vehicles.

During Dave's hearing concerning his termination by Fluor at Valdez, the following testimony was deemed irrelevant, so was stricken from the record. However, Dave considered the testimony relevant, because he had supposedly been terminated because of the costs involved from his alleged mistake. The testimony reveals the high costs caused by the high officials at Fluor. They are incidents which Dave witnessed. The testimony follows:

"Bobby H. had me put two tires in the back of the Chevy Blazer he was driving. This was obscured by visqueen. Bobby took out of the boxes in my tool shack a large assortment of various small tools, such as wrenches, pliers, screwdrivers, files, duct tape, etc. Bobby put these tools under his coat, always so that they were concealed. He always went with them to the parking lot where all the staff vehicles were parked.

"I have seen Bobby do this on several occasions and every time we brought any new tools up to the tool room from the warehouse, Bobby would check them over to see if there was anything he wanted for his own use.

"Fred helped himself to snow shovels, propane and other small items. On one occasion, we hid the shovels and the propane, as Fred ordered, under the snow in the back of his pick-up truck. I remember this incident very vividly because we hid these things so well that Fred didn't think we had done our job. Fred drove in and out through the gate with these things so well concealed that he didn't know the items were there.

"Fred came to me the next day and asked why I had not yet put these things in his truck. I walked over to said truck and pointed to where we had put the pilfered items under the snow. I told him the items were under the snow there where I pointed and that if he would just look when he got home he would find them.

"The first quarter inch electric drill on the job site was specifically for Bert, and I put the drill in his truck. The drill was never returned prior to my quitting the job. We also cleaned up, lubricated, and tuned up Bert's personal Johnson Sea Horse outboard motor in the laborer's tool shack.

"Bert's own personal boat was stored inside tank eleven at the east tank farm. This boat was cleaned by a crew of Chugach laborers.

"I personally did repair work on Harry's snow machine and personally witnessed other crafts doing repair work on Harry's snow machine. The carbide files Harry had welded on his skis were Alyeska property requested specifically by him. Harry, on numerous occasions, got small items like bug dope, WD 40, oil, etc., to take home.

"About twice a week we made a kitchen run for Fluor's office and staff at Power Vapor. During these runs we procured steaks, prime rib, pastries, cases of crackers and hot chocolate, cases of juice, etc. This food was eaten by Fluor staff on the Power Vapor site. The food was taken from the working men's kitchen in direct violation of Alyeska rules. This food was for the working men alone. Staff lives and eats at Fluor housing and does not have legal access to the working men's food supplies or kitchens. Yet they consumed large quantities of pilfered food."

Murphy's Law contends that anything that is not supposed to screw up can. Events like Love Canal, Three Mile Island, Chernobyl, Bhopal, and the Exxon Valdez oil spill would tend to lend support to Murphy's Law.

Dave's last job on the pipeline was at Coldfoot Camp, where he operated a John Deere loader on a traveling crew whose job it was to defer Murphy's Law for as long as possible. The perma frost under the pipeline was thawing, causing the vertical support members to slowly sink into the ground, which, of course, was causing the forty-eight-inch pipe to slowly settle. This would eventually cause cracks or breaks in the pipe welds, which would mean oil spills.

Alyeska decided to solve the problem by insulating the ground under the pipe with styrofoam or urethane. Dave worked on a small crew with three laborers, one of whom acted as foreman. Dave's loader was a small track-mounted loader with no cab and no roll bars. Every day the temperature was −45 to −48 degrees, with no wind. The dry shack was a Chevrolet Suburban that was usually gone when Dave needed to get warm.

The laborers were placing four by eight sheets of three-inch thick styrofoam on the ground in the affected area, which ran for miles. Two layers of this plastic insulation covered the ground for a total of six inches of plastic styrofoam. A dump truck brought sand, and Dave covered the six inches of styrofoam with eighteen inches of sand, squeezing under the forty-eight inch pipe with his loader. This type of insulation was performed on many miles of pipe. It was one of the most useless jobs Dave performed during the pipeline construction. If the pipe was settling during construction, with no weight or heat from the oil because the pipe was empty, what would happen after twenty years of hot oil passing through the line?

Dave believes there have been many oil spills which were covered up during the pipeline construction, and he is sure there are still spills no one hears about.

In spite of the sloppy, inefficient, mind-boggling manner in which the pipeline was built, it did get built. Oil has been flowing through it for thirteen years. The pipeline workers, known in Anchorage for throwing one hundred dollar bills around, frequenting the topless and bottomless bars, getting into brawls, and picking up women, disappeared nearly overnight. Anchorage was left with an oil-boom legacy of run-down massage parlors, displaced drunken natives, some fat wallets and some empty pocketbooks, and the prediction of doom. Truly local residents braced themselves for the inevitable depression. Alaskans were used to the boom-bust economy. They had lived through it with the gold rush, and with the military build-up. This was just one more bust.

But with the oil flowing through the pipeline, a new boom became evident. The oil companies own seven-eighths of the oil they can produce. The remaining one-eighth goes to the state of Alaska as its royalty share. With the royalties piling up in the state's treasury, the bust never really showed up. It was at this time that the people of the state truly began to believe the oil wealth would never disappear.

But the population boom had taken, and continues to take, a severe toll on the wildlife and the surrounding habitat.

One of the huge caribou herds of the state, which numbered in the thousands, migrated between the Denali Highway and the Glenn Highway. The season was open from August or September through March. The limit was three animals per hunter. But, as the population grew, the caribou were slaughtered by civilians and military personnel alike. They were slaughtered at Paxton on the Denali Highway. They were slaughtered at Eureka on the Glenn Highway. They were slaughtered at Lake Louise. Soon, the hunting was limited to a permit-only hunt, with very few permits issued.

With hunting curtailed, fishing became even more popular. During the sockeye run in the Russian River, it didn't matter any more if an angler went during the week instead of the weekends. It was still elbow-to-elbow fishing, known locally as "combat fishing." Fishermen have to cast and drift in synchronized rhythm or else risk a tongue lashing from a neighbor fisherman aggravated by the tangle of lines. Fights became common as the fishing grounds became crowded.

The Kenai River, overlooked in the 1970's as a legitimate salmon-fishing stream because of the cloudy water, soon exploded with fishing boats and guides, as people learned that not only was it possible to catch kings in the silty waters, it was exciting, and one king could provide almost all the salmon a family could want. But the wilderness experience was quickly dashed by boat traffic jams, with high-speed jet boats flashing over people's lines, cutting them, swamping smaller boats and eroding the riverbanks. Unchecked development along the river put docks and "erosion-control" projects in conflict with the salmon habitat. The edges of the river, where uneven banks and downed trees usually slowed the flow of the river and insured salmon fry could grow peacefully, were lined with cement walls—destroying the very reason people moved to the river's edge.

It was people—brought by the boom—who contributed to the destruction of the wilderness experience, and who may ultimately cause the destruction of the world-class salmon stream.

Excess people bring greedy entrepreneurs, and the Kenai River reached its peak of flagrant commercialism when an outboard-powered barge with a house on top of it was launched, advertising itself as a fast-food burger joint ready to serve hungry fishermen.

The high-speed boats and the burger barge were soon removed by an angry public, but the ravages of too many people continue to deteriorate the fragile river environment.

While it may be hard to believe that 250,000 people constitute too many people for the vast land area Alaska boasts, the problem is that those people's recreation is concentrated along two road systems, one which travels south to the Kenai Peninsula, and one which travels north toward Palmer. While there are a lot of float planes buzzing around the Alaskan skies, the reality is that the majority of the people can't afford flying. They must rely on family cars to get to the hunting and fishing. Add to the local residents scrambling to get out of Anchorage in the summer the tourists who flood the highways in their recreational vehicles, and the roadside parks are literally overflowing.

During Dave's first years of living in Alaska, he enjoyed taking the family on a Sunday afternoon drive to Wasilla, about fifty miles north of Anchorage. It was a pretty town, with a slow, relaxed pace. It had a post office, a couple of restaurants and bars, a store, the railroad, Wasilla Lake, Lake Lucille, and not much else. Once the New Parks Highway and the oil boom converged, that pretty little town was destroyed. As time went on, it turned into probably the world's longest shopping center.

At first, Dave and Betty used to take all five kids, load them into their International Travelall, and go on picnics, daylong fishing trips, and weekend family camping trips. They went to Peters Creek, Talkeetna, to the Russian River and Lower Russian Lake, to Eklutna Lake, to Lake Louise, to Hope, to Homer, and to any interesting place they could reach in their time off. But, as the crowds increased, so did the filth they had to encounter. The last time they visited Peters Creek, Dave had to take a shovel and bury people dung before his family could even pitch a tent and set up camp.

In these later years, a hike into Lower Russian Lake was the family's biggest disappointment. The lake is a clear mountain lake in a pristine setting. The famous Russian River runs out of the lower end of the lake, and there's a fairly easy four-mile hike to reach the cabin, maintained by the United States Forest Service, on the banks of the lake.

When the Brown family reached the cabin, they found it in shambles. There were beer bottles and cans strewn all over the lake shore, with numerous large pieces of visqueen.

The inside of the cabin was totally ransacked. Bears, although numerous along the salmon stream, couldn't be blamed. It was people. Ketchup bottles and mustard jars had been smashed against the walls, and some nice person had left a big pile of people dung right in the middle of the floor—complete with toilet paper.

During this period, Dave grudgingly spent a lot of his leisure time in the wilderness cleaning up other people's trash. On canoe trips through the Swanson River and Moose River, he would haul out beer cans, jars and paper. He could never understand why, if a person can carry canned food or jars of food on his back or in his canoe when they are full, why he could not also carry them back out when they were empty? The managers of the Swanson River Canoe System, the United States Fish and Wildlife Service, would have to recruit volunteers every year to clean up after those wanting a "wilderness" experience, and would end up hauling out many canoes full of trash.

Hunting practices also degenerated with the influx of people. Maybe some of the careless practices occurred in the pre-oil days because some people, particularly the kinds of individuals who settled in Alaska, would flout the law. But more people brought more law-breakers, and more problems to wildlife. One cruel habit by people who may call themselves hunters, but really are not, is the running down of moose and caribou by snow machine.

And though it may be illegal to hunt on the same day as a hunter has sighted his prey from an airplane, it still happens, shrinking to near zero any chance the game may have of avoiding the "hunter."

Before the law changed, Dave knew a bush pilot who, during moose season, would take a client in his Super Cub on floats, land next to a moose, kill the moose and take the client back by noon, pick up another and do the same during the afternoon. Each client paid two hundred fifty dollars for the privilege of shooting a moose.

While some of these practices have diminished, it is still common during moose season to see a plane circling a particular location, its pilot signaling to buddies on the ground where a moose may be located.

Poaching of game and snagging of fish is another serious

problem, and is almost always committed by people with plenty of money to buy meat from the local supermarket.

The city of Kenai boasts of a small caribou herd transplanted there several years ago in an effort to reestablish a herd which was devastated by wildfire and over-hunting. The herd is quite tame, and is a major tourist attraction. Local residents wanting to impress relatives can easily track down some of the herd members, including large bulls, for picture taking. One of the bulls was a favorite. It would wander close to the road and allow people to walk right up to it and get fantastic close-ups. An oil field worker Dave heard, who lived out of state, walked up to that bull with a gun instead of a camera, and killed it, taking only the antlers.

The state itself has been remiss in not making stronger laws amid the burgeoning population explosion. To hunt deer in Kodiak, Dave would have to take the state ferry system, riding the Tustumena Ferry to Kodiak. Several times he witnessed crew members dumping large plastic garbage cans over the side of the ships into the sea.

The state parks system, particularly on the Kanai Peninsula, has become a symbol of the skewed priorities of the state of Alaska, as latrines overflow, trash piles up, and recreational vehicles jam into and between unmarked camp spots. While the state spends plenty of money advertising itself as a tourism mecca, money was taken from the state park's budget and the system, at one time, couldn't even afford garbage pickup.

As Dave would enter one of the stinking outhouses overburdened with human waste, he couldn't help thinking about then-Governor Bill Sheffield's project, which used $2,700,000 to remodel the Governor's mansion. That money would have gone far toward establishing better rest room facilities and more frequent garbage pickup.

The state parks system tries to cope with the crisis by using volunteer labor for a lot of its projects, but still, the lack of funding indicates the perverted priorities of the Alaskan Legislature.

5

NEW FRONTIERS

Despite Dave's growing awareness of Alaska's environmental problems, these were the good years for the Browns. They enjoyed their comfortable house on Rabbit Creek Road on the outskirts of Anchorage. During the time the Browns lived in Anchorage, they watched it grow from 34,000 people, with dirt roads, rugged trails and friendly neighbors, to 250,000 people, with a McDonald's on nearly every corner. Once again, the notion of space, along with the desire for hunting and fishing, created an itch in Dave Brown's skin that couldn't be satisfied in the hustle and bustle of a city with the growing pains of Anchorage. The Kenai Peninsula began to beckon. And there was a business—waste disposal—with just the right lure, the promise of cleaning up Alaska, good profit and superb hunting and fishing. Dave Brown snapped eagerly at the bait.

The year was 1980. After fourteen years of hard work, Dave and Betty were seriously considering leaving Anchorage for a more leisurely way of life. They thought about moving to a little condominium they had in Whittier on Prince William Sound. Whittier is a beautiful place when the sun is out. The mountains, with their glaciers, slide straight down to the sea. It's the perfect disembarking point for exploring the wilds of Prince William Sound. But it can be quite gray and dreary during the many days of rain and overcast.

Then they considered moving to the five acres they owned on the Susitna River. They'd have salmon fishing at their rod-tips, and moose hunting all around. They couldn't decide. Then, one spring day, Dave went to a boat show. He dashed home to tell Betty.

"It's absolutely perfect!" he confided to his wife. "I've found a twenty-three foot Correct Craft Fish Nautique fiberglass inboard, with a center steering console, canvas, a windshield, and a three-hundred-fifty-one horsepower Ford engine. They had Correct Crafts when I was a boy on Lake Ontario. This is absolutely the finest fishing platform ever!"

A little non-plussed, but ever supportive of Dave's enthusiasms, Betty replied, "Well, why not? Let's take the boat to Whittier and live in our condo."

However, Dave knew living in the condominium in a town as tight and small as Whittier might prove to be fairly restricting. Whittier's only access is by train and by boat, and most of the town's residents live in one apartment building.

He tried to entice Betty further, "We could also build a cabin on our Susitna land and go there whenever we want to get away from Whittier. Shoot! We could even head south for the winter if we wanted to."

So it was settled. Dave decided to enjoy life as a fishing guide, and maybe take a few hunters out in the fall. But plans change, and sometimes the lure of exciting challenges and big profits can cause people to make huge mistakes.

When the Browns decided to move, Dave was working as a real estate salesman. He had never made a fortune but he had a listing for two gold claims near Talkeetna. Two men from the Kenai Peninsula bought them, giving Dave an earnest money payment. The owner of the claims, Tim Geiermann, wanted the rest of the money; so he gave Dave a round trip plane ticket to Kenai to collect it from them.

When Dave walked into the coffee shop of the Kenai airport, he saw three, not two, men waiting for him.

"I'm Dave Brown," he said, quickly introducing himself. And then, in his usual forthright manner, he asked, pointing his finger, "Who's he?"

"Well, these two youngsters want to get into the gold mining business," the slick, black-haired man, whose name was Tom Lyan, laughed. "They've got raw muscle and guts, but not much in the way of money. So, since I have a lot of it, I'm going to provide the cash."

Dave's curiosity and distrust of bragsters got the best of him.

"Well, Tom, where'd you get all your money?"

Tom looked searchingly at Dave and answered in a voice which Dave later described as "butter would melt in his mouth" and smiled again.

"Well, you see, the oil industry is a big business down here, and you just got to learn how to tag on to the oil companies' coat tails. I own a business called Mar, Inc. I have four vacuum trucks and operate the Sterling Special Waste Site.

Tom dangled the bait, "The set-up is a money tree."

Dave began to listen more intently.

"Sounds first rate," he said a bit jealously. "You're lucky."

The man nodded, watching Dave and sensing a bite on his line.

"My site is the only one on the peninsula where the oil companies can legally dump their drilling muds. I operate the site for the borough, and the borough guarantees me that the muds can be dumped nowhere else. And the oil companies are always drilling new wells, so there's plenty of business. I work from 8 a.m. to 5 p.m. Monday through Friday. It's a good life, but I've made my pile and I'm going to retire one day soon, move closer to my kids, and sell the business."

"You are?" Dave said, by now excited.

"Yes, Alaska has been real good to me," Tom said slowly, watching Dave all the time. "I've got this cedar home right on the banks of Longmere Lake and can catch rainbow trout and

Kokanee salmon not more than fifty feet from my front door. And, of course, there's always the Kenai River just ten minutes away, where I can catch kings, reds and silver salmon, along with rainbows and Dolly Varden. And if you like the quieter sport of exploring the wilderness, there's the Swanson River Road with its canoe trail system, and ducks, geese, loons, trumpeter swans, rabbits, spruce hens, bears and lots of moose."

By now, Dave was ecstatic. Everything he ever really wanted in life appeared to have come right to his doorstep. A chance to do something he believed in—clean up Alaska, live the life he'd always wanted—unencumbered hunting and fishing, and make a profit to boot.

"Boy, if you ever put a price on that business, I sure would like to bid."

Tom Lyan looked off into the distance. "When I do, I'll be glad to contact you," he said. "After all, we're already friends, aren't we?"

After collecting the money for the mining claims, and indicating an interest in Mar Enterprises, Brown flew back to Anchorage.

Not long afterward, Tom Lyan called.

"Look, I've made my final decision and I'm giving you first chance. I know how bad you want it, and we're friends, aren't we?"

Dave quickly phoned Ray O'Docharty from Wasilla to see if he was interested in entering into this venture. He was, so the deal was quickly settled. Neither Brown nor O'Docharty ever seriously questioned Tom Lyan's honesty. Dave had been enticed, fallen hard, and felt he could trust Lyan. After all, Lyan was an Alaskan who liked to do the same things Dave did and who appeared to have the same goals in life: hunting, fishing and enjoying the good life of the out-of-doors.

On July 7, 1980, Brown and O'Docharty completed the purchase of the business from Lyan. Dave and Betty went on to purchase Lyan's home on Longmere Lake. They couldn't wait to begin enjoying the good life.

It didn't take long, however, for Dave to smell something totally foul in the business deal. Dave and Ray began to suspect Lyan had only told them part of the story. He had played show and tell with Dave and Ray, and the Peninsula borough government, showing them what he wanted them to see and withholding from them things they would find out later.

The biggest area of deception was that Lyan had to have known the Sterling Special Waste Site was doomed. While it was true that borough engineers recommended the site because of its number of feet above the water table, it was also true that the site was located in the backyard of too many people. Some local residents were pressuring the borough hard to close the site, located on a hill within a mile and a half of the small tourist community of Sterling. Among them were Walt and Elsa Pedersen. The Pedersen's verbalized the thoughts and fears of other people of Sterling. They constantly wrote to the newspapers, the Alaska Department of Environmental Conservation (DEC)—which was in charge of permitting the site—Alaska legislators, Alaska's United States Senators and its single Representative.

Dave and Ray didn't know it yet, but the residents' fears were not unfounded.

The Sterling Special Waste Site was established by the Kenai Peninsula Borough at the request of the state. The state deeded over the land to the borough, so the borough would then be responsible for administering the site. Oil had been discovered in 1957 on the Kenai Peninsula in the part of the Kenai National Wildlife Refuge now known as the Swanson River Oil Field, and drilling had gone on virtually unregulated until 1973. After that the newly established Alaska Department of Environmental Conservation came up with regulations making it illegal to dispose of solid waste, including drilling muds, without a permit. The Sterling Special Waste Site was to be the collection point for all those oil field wastes, which not only included the bentonite clays used for drilling muds, but everything else that goes down the hole, including acids, caustics, corrosion inhibiters, and hydrocarbons—chemicals routinely dumped in the drilling mud reserve pit along with the cuttings.

The borough was deliberate in its selection of a site. It

wanted a happy oil industry, yet the oil industry needed to comply with the new solid waste regulations. For three years, engineers studied the land in the borough looking for a site. The Sterling site was chosen because it was one hundred feet above the water table.

The site, permitted through the DEC, was allowed to accept three kinds of waste: water containing less than four percent oil, except emulsions, which could exceed four percent; solid waste contaminated with petroleum and its products (drilling muds); and non-toxic water conditioning sludges which normally came from a Unocal Chemicals fertilizer plant in nearly Nikiski. The site would also be allowed to accept other wastes "only after prior written permission had been obtained from the Alaska Department of Environmental Conservation." No hazardous wastes were to be accepted. Although drilling muds often contained such hazardous materials as acids, phenols, and heavy metals [See Appendix A], the oil industry had successfully lobbied against including its industries' drilling wastes in the United States Environmental Protection Agency's list of hazardous wastes. As a result, most of the wastes were buried on site.

The original intent in establishing the Sterling Special Waste Site was a good one. The solid waste regulations imposed by the fledgling DEC were stringent ones, making it difficult or nearly impossible to obtain a drilling mud disposal permit on the Kenai. The regulations stated that the distance between the pit bottoms and the water table had to be at least six feet. Most of the Kenai Peninsula is wetlands, precluding the legal disposal of drilling muds almost anywhere the oil companies were drilling. Although the regulations went into effect in 1974, there was no legal place for the oil companies to put their wastes. They continued their practice of burying them in unlined pits at the drill sites.

With the establishment of the Sterling Special Waste Site, not only would the oil companies be able to then obey the law, it would be easier to monitor one site instead of trying to monitor all the drilling sites.

On October 6, 1975, the Kenai Peninsula Borough Planning and Zoning Commission had met in Soldotna. Testimony was given at that meeting by George Day, at that time manager of the Chevron Oil Company Refinery in Nikiski, which is the Kenai Peninsula site of four different refineries. Day was representing the Alaska Oil and Gas Association (A.O.G.A.). Day read a statement from the association urging early approval of the Sterling Special Waste Site. He said that because of changes in statutes and regulations, the oil companies could no longer dispose of drilling mud on site, they could no longer take drilling mud to the Kenai Landfill, and shipping drilling mud outside to approved sites in the Lower Forty-eight was cost-prohibitive. In researching the history of the site later, Brown had to chuckle over that testimony, because there was not one scrap of evidence that Chevron disposed of any drilling muds at the Sterling site while Lyan operated it, and the company didn't dispose of muds while Brown operated it, despite the fact the company drilled thirty-one wells near the site, from the time the site was opened until 1986. Of course, that company wasn't the only one.

Convenience was another reason the site was selected. It was located just twelve miles down the road from the Swanson River Oilfield. Trucks hauling the drilling muds from the field could go directly to the Sterling Special Waste Site without traveling through any residential areas, or even having to use a highway. That, plus the desire to keep the oil companies happy, plus the fact that on maps it was located one hundred feet above the water table, contributed toward the decision to go ahead with the site.

While the procedure for selecting the site was long and tedious, once selected, the site was ill-prepared. It had long been used as an unofficial landfill. The site was filled with old car bodies, truck frames, and assorted other junk, which was then covered with dirt. The waste pits were then excavated in this fill dirt, so that regardless of the supposed good soil conditions, the dirt could not be compacted, making the potential for leaching inevitable.

That leaching showed up quickly. Although no one had told Ray or Dave, phenols, heavy metals and oil were found in

1976 in samples from four monitoring wells. The DEC consistently claimed, however, that the samples were no good, having been taken from a pool of water in a galvanized pipe, instead of using a legitimate sampling method. However, according to a Water Sample Analyses Results Summary, the contaminants were found in four different monitoring wells at the site. In 1980 and 1981, only a few of the wells, different from the 1976 wells, were tested. Much less contamination was recorded in those wells.

The Sterling site was operated by the Kenai Peninsula Borough from 1976 until September 1, 1978, when it was turned over to Tom Lyan. During these two and a half years, the borough was required to keep records including the quantity and type of waste received, the source of the waste, and water quality analysis from samples taken from monitoring wells.

Brown requested these records from the borough after having operated the site, because, as site operator, he thought he might someday be held liable for the disposal of unpermitted waste by either the borough or Lyan. He was told the borough had no such records.

To Dave, in addition to the leaching, the site had been operated haphazardly. A letter to the DEC from the engineering firm, Dowling, Rice and Associates, showed that erosion of a dike embankment used to hold muds generated by Union Oil had taken place, and despite the firm's assurance that no drilling muds had escaped, aerial photos supplied by the firm showed drilling mud in the woods outside the fence, and way outside the site boundaries.

It's possible that when the engineers inspected the site, the mud was still where it was supposed to be. Perhaps the most extensive damage occurred later. Or perhaps Lyan hadn't told Mr. Rice, the engineer with the firm, what was happening.

Lyan had told Ray and Dave he wanted to sell the business, because he had made his money, didn't need the hassle any more and was worried about his heart. The truth was that his business had fallen way off, although he and his accountant didn't disclose that fact to Brown and O'Docharty. According to other reports which Dave and Ray heard, the Sterling Special Waste Site had also been operated recklessly and irresponsibly with little regard for the law or common sense. Lyan was about to lose his permit to operate the site, and he knew it.

In July of 1980, Richard Williams, an environmental engineer with the DEC in Juneau turned in an inspection report of his visit to three Kenai Peninsula solid waste facilities. Concerning the Sterling site, he said:

> "This facility is inoperable in its present state. The ponds are all heavily oil contaminated and it appears that other wastes have been placed improperly. I recommend the permittee be directed to cease operation immediately using an emergency order, if necessary. The wastes in the site should be sampled and checked to insure that no hazardous materials such as PCB's have been deposited here.
>
> "A compliance order for the cleanup of the site should then be issued. This should also include the installation of a proper ground water monitoring system for use with or without future site operation. The permittee and the owner should be given the option of site closure and restoration or resuming operation under strict enforcement of the permit.
>
> "Cleanup should be completed not later than October 1, 1980, according to plans submitted to and approved by the department.
>
> "The facility's users must be notified of the closure and assisted in finding alternative disposal means. This should not be too difficult as the site does not appear to have had significant use recently."

On July 7, 1980, the very day that Ray O'Docharty and Dave Brown purchased Mar Enterprises from Mr. & Mrs. Lyan, Mr. Williams of the DEC recommended that the site be closed.

But Brown had no idea of the cloud hanging over his new business from the start. When Ray and Dave started work on the site, they had high ideals of running a legal and clean dump site and they relished the challenge. Unfortunately, what

Dave and Ray discovered when they drove around the south end of the property, escalated the challenge into a crisis.

Dave called out, "Look at that!"

"What?" Ray asked, unsuspecting.

"Those two huge trenches," Dave spat out. "They're filled with drilling mud and dirty water."

On closer inspection, the two men found the water was too deep to evaporate, and the drilling mud prevented it from leaching into the ground quickly enough.

"We'll have to dewater those pits," Dave said angrily, deciding he would have to dig new pits adjacent to them and pump the water into the new, clean-bottomed pits.

"The new pits, having no drilling mud in them will leach the water into the ground easily," Dave figured.

"We'll have to wait 'til next week," Ray said. "It's Friday afternoon."

Early that Monday morning, the two men brought in their machines and began to excavate two or three feet into the dirt.

"Dammit," Dave said to Ray, "Look at this shit."

He had encountered scrap iron, old truck frames and car bodies.

"I'm going to complain to the borough solid waste engineer."

At the borough office, Dave was told by the official in charge, "Oh, you haven't seen anything. You haven't scratched the surface. That stuff is all over under this site," he was told.

Dave couldn't believe what he had gotten into. To turn that location into a special waste site on top of that kind of debris was a dangerous and irresponsible act by both the state and the Kenai Peninsula Borough, laying them wide open for potential future liability.

Dave was beginning to get an idea of the corrupt nature of the oil business and disposal of its waste. And his awakening continued as Dave began to discover how Lyan had dealt with bureaucrats.

On September 1, 1978, the Kenai Peninsula Borough had entered a twenty-year lease with Mar Enterprises, owned by Lyan, and a year later agreed to sell to Mar Enterprises the existing capital improvements at the site. It was this second agreement which seemed to be the problem. In an affidavit to the borough used to put a value on the collateral Lyan used against the purchase of the site improvements, Lyan stated that he owned three tanks. Not surprisingly, the three tanks were valued at $39,000, just $487 less than the purchase price. There was an agreement that the borough could seize the tanks, and Lyan would have to pay the rest of the money if there were any capital losses at the site [Appendix B].

When Brown took over the site, the tanks did not exist. No ten thousand-gallon tank valued at $3,000 existed. The three thousand-barrel steel tank with an estimated value of $30,000 was pure junk. It was a tank that had been destroyed by a fire at Swanson River. Dave heard it was given free to Lyan for dismantling and removing it. It lay on top of the ground at the Sterling site in three piles of rusty, bent, bolted-together panels with two fifty-five-gallon barrels full of rusty nuts and bolts with which to assemble it.

In the files of Mar Enterprises, Dave found a hand written quote from F.R. (Peanut) Melott of Oil Industries in Soldotna to erect that same three thousand barrel tank and put a liner in it, so it would not leak. After having been through a fire, the tank would leak like a sieve without a liner. Peanut's entire quote to refurbish the tank came to $20,817.62, as opposed to a new tank costing $16,175 F.O.B. Tulsa, Oklahoma, without a liner, plus freight on 23,800 pounds and erection costs.

As Dave was going through the Mar files, he wondered how the Kenai Peninsula Borough could not notice the truth. The borough had a person who visited the site on behalf of the borough to protect the borough's interest. This man had a sign in his office stating, "Engineer, Faultless and Perfect," so Dave began referring to him as Mr. Faultless and Perfect. Getting to know Mr. Faultless and Perfect helped Dave theorize how oversights by the borough and the state could come about.

Mr. Faultless and Perfect often traveled to the Sterling Special Waste Site with a man from the DEC who Dave called

John Dough. Dave gave him the name of John Dough because he was broke in early 1982, but before the year was over he was flaunting his wealth, and people wondered where it came from. Dave referred to John Dough and Mr. Perfect and Faultless as the "Gold Dust Twins."

On one occasion they approached Ray and Dave.

"We're going to Homer halibut fishing in the next day or two and we're intending to take your new twenty-three-foot Correct Craft boat and trailer."

They didn't ask Ray and Dave if they were going fishing, and could they take them along. They didn't even ask if they could borrow the boat.

Dave's temper flared. "The hell you will! The replacement cost of that boat is about $25,000 and you couldn't reimburse us if something happens."

Ray added gamely, "Besides, it's a double conflict of interest, with both the state and the borough."

But that wasn't the end of asking for favors. On two other occasions, Mr. Faultless and Perfect tried to get Ray and Dave to bankroll or back an invention or design of his. His final attempt was a plea for the two of them to set him up in the private sector. He asked the two of them to establish his business and provide him with a salary for a year or two, then, of course, be his first paying client. Ray and Dave consistently turned down the requests for favors, and Dave believes that may have been the beginning of his problems both with the state and the borough.

In mid-1981, after Brown had been in business about a year, Mr. Faultless and Perfect came to his shop building in Soldotna and approached Ray and Dave, in the presence of Dave's wife, with a strange request.

"I want to give the two of you some money," he said slowly.

Dave's eyebrows raised.

"Tom Lyan has given me two septic tanks which I've installed at my home. From our previous encounters, I've seen you two won't take or hand out favors, but this time I owe you."

Dave speculated that he must have been afraid that since the tanks were once the property of Mar Enterprises that perhaps he'd better pay for them even though it was a year or two after the fact, just to preclude any potential conflict of interest.

Dave told Mr. Faultless and Perfect to go home and forget it, that the transaction had occurred before their time and it was none of their business. He made it clear that there was no record of the gift of the tanks and that they did not want his money. Finally, the engineer reluctantly left.

A few days later, he returned, this time trying to trade a twenty-five horsepower motor for the tanks. Dave and Ray again sent him home telling him to forget it, that they did not want his money or his motor or anything else, that the matter was over and should be forgotten. Faultless and Perfect came back a third time. This time he sat down and wrote a check to Mar Enterprises for four hundred dollars for two used septic tanks. Dave was tired of seeing his face and hearing him whine, so he took the money.

Brown never found any proof that Mr. Faultless and Perfect was responsible for the borough's oversight in the matter of the worthless tanks. But considering how he and Ray were treated during the early take-over of the site, he couldn't help wonder if a person who appeared to know as much as Mr. Faultless and Perfect could overlook the fact that Lyan's tanks were non-existent. Dave then began to wonder if there were other "oversights."

John Dough also continued to ask for favors. Only a month after Ray and Dave purchased the boat, and too early for John Dough to get acquainted with them, Dough came down to the dock site.

"I really like your boat," he said admiringly. "And since you have two other boats, I propose that you and Ray get into the oil spill cleanup business."

He didn't wait for an answer. "For a mere $35,000 a year, with my connections, I can almost guarantee that Mar Enterprises will get to clean up all the oil spills on the Kenai Peninsula and in the lower Cook Inlet."

"I don't know," Dave said slowly.

Dough continued, "Look, all I want for now is to be able to use the boats whenever I want. Of course, when business picks up, you'll be able to pay me accordingly."

Dave and Ray passed a knowing look between them and Dave answered tight-lipped, "No, I don't think so. We just bought a new business we really don't know very much about, and we don't want to get strung out any further."

Later, when Dave and Ray became involved in an oil spill clean-ups, Dough approached them again. This time he seemed even more slippery than usual.

"I just want to advise you which company you should buy your oil spill clean-up materials from, and the individual in the company you should deal with," he said with a smile.

Dave's grim-faced, fast reply cut him off.

"It isn't your damned business to recommend particular companies," Dave said quickly walking away.

But the incident stuck in his mind.

6

BENEATH THE SLICK SURFACE

After Dave took over the operation of the Sterling Special Waste Site, it soon became apparent that the wastes to which he was entitled—all the drilling wastes generated on the Kenai Peninsula—were not coming to the site. In fact, in the beginning, he made just enough money from the site to live comfortably; he did not make what most would consider a reasonable profit. He also noticed that the amount of drilling wastes also began to dwindle as he refused to cooperate with what he considered the unethical behavior of borough and state officials.

But his gut feeling had to be substantiated. So Brown undertook a research project, looking into just how much drilling wastes should have been going into the site.

Each time a well is drilled, whether it is offshore or onshore, waste is generated. The major portion of that waste is drilling mud.

With the DEC's regulations in 1973, requiring a permit

for the disposal of solid waste, there were certain safeguards that were to be met in order to get the permit.

According to the DEC, drill muds mean the fluids composed of clays, oil, water and chemical additives used by oil wells and other deep drilling operations to lubricate the drilling bit, maintain hydrostatic pressure control in the well, remove drill cuttings from the well and stabilize the walls of the well during drilling.

The oil industry would like the public to believe that drilling mud is harmless, because its principle ingredient is bentonite clay, a pure product of mother nature. During many public hearings on the Kenai Peninsula concerning permits to dispose of the muds, and during hearings in Anchorage concerning the reconsideration of drilling muds as hazardous wastes by the Environmental Protection Agency, industry officials continued to maintain the muds are mostly harmless, and that any additives are diluted beyond the necessity of warranting special handling.

Bentonite clay by itself is harmless. The harm comes from the contaminants that frequently, but not always, appear in the finished product, the used drilling mud. These contaminants appear for various reasons. First, there are the chemical additives that each oil company or mud company puts in its drilling mud in order to get the job done. For the most part, these are proprietary secret formulas. They use materials to fluff the mud up. They use aerators and biocides and elements to drop out the oil. They use corrosives and scale inhibiters to prevent corrosion and scaling of the pipes. They use acids to acidize the rock to help get out the oil. Most oil companies consider the ingredients they use in their muds to be restricted information. Some of the additives are harmless. Many are not [See Appendix A].

Some of the ingredients in used drilling mud are not chemical additives but instead are properties picked up in the course of drilling that exist naturally in the earth itself prior to the arrival of the drill bit. In drilling, different zones are encountered, and different zones naturally produce different products. Heavy metals and arsenic occur commonly and naturally in the earth and are disturbed in drilling, frequently ending up in drilling mud. Sometimes a certain amount of crude oil appears in the drilling mud as well as condensate which contains aromatic hydrocarbons, such as benzene, that are soluble in water.

At times diesel is injected into drilling mud for lubrication or to prevent it from freezing. Sometimes other lubricants are used. A major and serious consideration with used drilling mud is that reserve pits on drilling sites are frequently used as garbage dumps, and all of the trash, debris, and unused or used chemicals end up out of sight in the bottom of the pit. So while the muds might start out harmless, by the time they have run their entire course, they may be harmless no longer.

The main reason for getting a disposal permit is to see that the proper safeguards exist to ensure that no pollution of the land and water will take place. Alaska state solid waste management regulations state, "A person who constructs, modifies, or operates a solid waste disposal site shall do so in accordance with a waste disposal permit issued to that person by the department."

Some may not consider a regulation to be as important as state statutes. But solid waste disposal was addressed in Alaska Statutes as well: "Waste Disposal Permit Procedure: An application for a permit shall be made on forms prescribed by the department or on forms prescribed by the United States Environmental Protection Agency and shall contain the name and address of the applicant, a description of his operation, the quantity and type of material sought to be disposed of, the proposed method of disposal, and any other information considered necessary by the department."

A permit is not merely a piece of paper designed to provide work for a clerk typist. It is a document designed to protect the environment from pollution. To apply for that permit, one must first prove that his disposal pit bottoms are at least six feet above the water table. He must prove that his pits are at least one hundred feet from surface water of any kind. The applicant must furnish the landowner's written permission. Then there are specific stipulations concerning the individual site, such as type of liners, type of restoration when the pit is finished being used, and type of monitor for potential leaching. The permit is to ensure safe disposal. No one is supposed to be able to dispose of solid waste without a permit. If the permit application conditions cannot be met, it is mandated that there will be no disposal at that particular location, because it wouldn't be environmentally sound. The permit is designed to protect the environment and the people from poisoning.

In his research, Brown had discovered that the state of Alaska had totally ignored its own regulation which says, "A person who owns or operates a property shall remove accumulated solid waste to a permitted disposal site." As many as one thousand seven hundred of the unpermitted drilling mud disposal sites are on state-owned land.

Even the oil industry, in its 1976 support of the Sterling Special Waste Site, declared that the new law prohibited them from disposing of the muds in the traditional pit at the drill site.

Further, Brown discovered that the industries' support of the site was in word only, and not in deed. It appeared that although the industry officials must have realized burying the wastes was illegal, it was much cheaper than hauling them to the special site. Yet, at the same time, the public was beginning to get worried about the potential contamination of the ground water. Having a site exist, lulling the public into thinking all the waste was being taken to one spot, would be ideal for the oil companies, and all they would have to do is use it once in a while to keep everybody happy. The Sterling site was an alibi for oil company polluters who feared they someday might need an alibi. Company officials probably had no idea someone would actually calculate how much mud should go to the site and how much actually got there.

When Brown took over the site in 1980, he began keeping accurate records. When he realized he wasn't getting the drilling muds he needed to make an adequate living, he shared his information with DEC Kenai field officer Howard Keiser.

In an October 20, 1981, memo Keiser wrote for DEC files, sending a carbon copy to Bob Martin, then special assistant to the DEC commissioner, and Ray Norris of the EPA, he stated, "It appears that by the Sterling site being in the news and due to the resident's [sic] objections, that a more serious pollution problem may exist regarding drilling mud. Drilling mud at one time used to go to Sterling. It is now buried on site at the drilling rigs, often in wetlands or in areas of high water tables. This is based on my observations and records of Mar Enterprises. The last time any drilling mud was taken to Sterling was August 14, 1981, yet we have had nine new wells drilled in this area."

Through his research, Brown accumulated complete lists of all wells drilled for oil and gas in Alaska, both onshore and offshore. In addition, he also had complete lists of all drilling mud disposal permits issued and applied for throughout the entire state, and sample application forms for drilling permits and disposal permits.

As he continued gathering material, he felt sure that the oil companies had frequently, flagrantly violated both their legal and moral obligations to the people of the state of Alaska and of the United States. He also felt that most of the dirty tricks that were played in the state of Alaska by the oil industry were played on state and federal lands.

The actions by the oil industry, and subsequent inaction by the DEC are in direct violation of the state's own statutes: "It is the policy of the state to conserve, improve and protect its natural resources and environment and control water, land and air pollutions in order to enhance the health, safety and welfare of the people of the state and their overall economic and social well being.

"It is the policy of the state to improve and coordinate the environmental plans, functions, powers, and programs of the state, in cooperation with the federal government, regions, local governments, other public and private organizations and concerned individuals, and to develop and manage the basic resources of water, land and air to the end that the state may fulfill its responsibility as trustee of the environment for the present and future generations."

Again, the Sterling Special Waste Site was established to

implement on the Kenai Peninsula that statute which read (Section 46.03.100) "A person who conducts an operation which results in the disposal of solid or liquid waste or heated process or cooling water into the waters or into the land of the state must procure a permit from the department before disposing."

Section 46.04.110 states: "Application for permit shall be made at least sixty days before commencement of a proposed discharge." Anyone who deviated from these sections was guilty of a criminal act. The Statutes were clear. Polluting the ground or water in the state of Alaska is a crime.

The Sterling Special Waste Site, regardless of its appropriateness as a waste site, held the only valid and legal permit for the disposal of drilling mud on the entire Kenai Peninsula when Dave took it over.

Later, a permit for difficult-to-pump mud was issued to ARCO. By the time ARCO began drilling Wolf Lake #3, there was a dark cloud hanging over the future of the Sterling Special Waste Site, and ARCO applied for and received a permit from DEC to dispose of their mud on site in a lined pit. It was the first legitimate lined pit to ever appear on the Kenai Peninsula.

Ninety-six wells were drilled onshore on the Kenai Peninsula. At nearly every well, all or most of the mud was buried unlawfully at the drilling site. In some cases, token quantities were brought to the Sterling site to give the oil industry the smoke screen it needed.

While Brown operated Mar Enterprises, Inc. and the Sterling site, forty-four wells were drilled onshore in the Kenai District. Of those wells, Brown received drilling mud from only nine locations. He received minimal quantities of drilling mud from three Unocal locations, including Cannery Loop #2 and #3, plus one well in Union's Kenai Gasfield. Out of seventeen wells drilled by Unocal while Brown operated the Sterling site, Dave believed Unocal partially adhered to the law on three occasions, and violated the Alaska Statutes and regulations a total of seventeen times.

Thirty-one wells were drilled in the Kenai District by

Chevron, with Chevron finally applying for and receiving two mud disposal permits in 1985, for wells drilled in the Swanson River oil field. The issuance of both of these permits raised serious questions, since one of them may violate the economic viability clause of the lease between the Kenai Peninsula Borough and Mar Enterprises for operation of the Sterling site . . . the clause which stated all drilling mud generated on Borough-owned lands was to go to the Sterling site. The other permit may violate the state's own regulations. The permit is for a disposal pit, the bottom of which is from four to six feet above the water table, in spite of the fact that state regulations clearly state that the pit bottom must be at least six feet above the water table. The regulations do not say four to six feet. That leaves, however, at least fourteen wells from which Chevron should have brought muds to the Sterling site but did not.

Marathon Oil drilled a well at Beaver Creek, completing it in September of 1982. The company brought quite a bit of mud to the Sterling site. The question is, did they dispose of the rest unlawfully by spreading it on the roads at Beaver Creek—in total violation of the Statutes of the state of Alaska? Brown was told by some of the workers at the Beaver Creek site that the spreading of the muds was done covertly.

There were two wildcat companies drilling during Brown's tenure at the site. He received small quantities of drilling mud from Amerex, but never got paid because the company filed for bankruptcy. The site also received mud from Far North Oil and Gas.

Legally, all the drilling mud from the forty four wells should have gone to the Sterling site. It wasn't the fact that the site was perfect. It wasn't. But it was better than having drilling muds scattered in the peninsula's wetlands, mostly on state and federal lands. As it stands today, drilling mud is disposed of at three hundred and fifteen separate onshore wells, with only five permits.

The sites at which oil companies disposed of wastes are remote, and, of course, not open to the general public. But being remote does not lessen their ability to pollute.

Marathon's Beaver Creek Field, with all of its disposal

sites, is located in the immediate vicinity of Beaver Creek aquifer, which supplies the water for the city of Kenai.

Ground water from the Swanson River field, which has been in operation by both ARCO and Chevron since 1957, flows directly toward the Beaver Creek aquifer. The ground water at Swanson River has also been proven to be contaminated.

7

PRICKLING CONSCIENCES

During the early years of oil development in Alaska, Dave Brown, like many others, did not pay too much attention to the resulting waste, because like them he wasn't aware of it. Everyone was revelling in the newfound wealth. Once again, the adventurous spirit that developed the mining of the west infused the pioneers of the last frontier. That spirit decreed it acceptable—even manly—to take what it could from the land, and not worry about the future.

But with the environmental movement of the 1970's, Dave, like some other people, started to think about the future, and what it might be like to live in this beautiful country with waste all around. When the Alaska Department of Environmental Conservation was formed in 1971—to give assurances to the people of the state that industry wouldn't rape the land, Dave and others who had come to love the land, breathed sighs of relief.

Still, while the idea was a good one, it didn't get the kind of financial backing it needed from the state. Little by little, legislators, supported in their campaigns by the oil industry, began to hack away at the DEC's budget until there was only one person in the entire state monitoring for clean air. There were only two field officers working on industrial matters on the Kenai Peninsula, where oil development was the oldest, and only two field officers for Fairbanks, which covered the North Slope.

While the DEC's duties included making sure the industry complied with the law, agency officials also performed such varied work as approving subdivision plats, checking for leaky underground storage tanks, inspecting restaurants, doing the paperwork on required monthly water testing of commercial buildings, responding to oil spills, and making sure Alaska's seafood is fresh. The department's budget was the smallest in the state, not counting the Department of Veterans and Military Affairs, which received mostly federal money. On top of that, just when Dave and other people were clamoring for better pollution control, the budget was cut by one-third.

But a small budget is not an excuse for not getting a job done. As Dave's awareness of pollution grew, he repeatedly began to contact the DEC with information about dumped drilling muds. He was repeatedly ignored, like a person would ignore a mosquito that was flying around, but not landing. Brown couldn't help but wonder if he were being ignored because the agency which was supposed to regulate the industry was actually in the pocket of that industry.

Dave's disillusionment with public officials burgeoned. He began to wonder how he was going to be able to get his message across. The information which Brown discovered about the lack of permits for disposal of drilling muds was readily available to anyone who wanted it. No one seemed to want it. The Alaska Oil and Gas Conservation Commission can provide a computer printout detailing the wells drilled in Alaska, both onshore and offshore. For each well listed there is the well name, the operator, the date spudded, the date completed, the section, township, range, meridian, lease type and the name of the landowner. Copies of each permit to drill are sent to the DEC [See Appendix C]. Yet when Brown requested from the DEC an updated list of drilling mud disposal permits, agency officials stated they couldn't supply Brown with the names of the landowners. Brown wrote back that he found that strange, since the law requires permission of the landowner before a disposal permit can be issued.

While Brown's accusations of oil company dumping were being ignored, the fact that Brown operated the Sterling site was not. Brown wondered if his refusal to enter into the game of favors contributed to this stringent policing of the tiny Sterling site while massive illegal dumping was ignored.

On the other hand, Dave—better than anyone—knew the Sterling site obviously had many problems. First, it was ill-prepared to accept waste, with its buried car bodies and scrap iron. Second, it was too close to people—people fighting to close it down. The ultimate goal of everyone involved with the site except Brown, as the operator, seemed to be to shut it down. But Brown felt everyone was missing the obvious. Waste had to be disposed of, and at least he was trying to find a way to do it legally, effectively, and cleanly. Still, there was a lot Dave Brown hadn't been told. Tom Lyan knew there were efforts to shut down the Sterling site before he sold Dave the business. The state of Alaska knew it before Brown took over the site operation. The Kenai Peninsula Borough knew it, and the people living in Sterling knew it. But instead of putting all the cards on the table, and saying here's what we want, Dave Brown was harrassed, lied to and a phony paper trail of alleged violations was created. Meanwhile, the rumors flew. Dave Brown appeared to be a Don Quixote, fighting windmills. No one took him seriously. And the oil companies wanted to make sure no one *ever* took him seriously.

Of course, although several members of the oil industry had originally wanted the site, they too now wanted it shut down. While Tom Lyan hadn't raised any questions about the amount of mud coming to the site, Brown had. The oil industry didn't want someone keeping track of how many wells, and how much mud was being illegally dumped. Even token use of the site was too expensive considering the cheap alternative of burying drilling mud on site, free of charge, and without the hassle of applying for a permit. They worked behind closed and locked gates, so they feared no interference from anyone.

The major cost to the oil industry in having the site remain open was that Brown was constantly aware of who was drilling and where. Dave knew who the violators were, and who the good guys were. He knew how big the reserve pits were, how deep they were and how much freeboard, if any, still existed on the sidewalls of the reserve pits. This told Brown approximately how much drilling mud was used and should be coming to Sterling. In addition, Brown had friends and sons of friends who worked in oil fields. His employees had friends who worked in the oil fields. Some of the men who worked in the oil fields did not like what they saw, but they needed their jobs. So, rather than complain to DEC, they came to Brown. And, as his knowledge grew and his vocalism increased, Brown became a liability that the oil industry could not afford to have around.

Somehow the focus of concern about oil dumping suddenly became not the big offenders, but Dave's site. Rumors were started and misfortune played into the industry's hand when an erosion at the dike at the Sterling site caused a spill outside the site perimeter. This breach in the dike convinced the people of Sterling that Love Canal could happen in Sterling. Although Brown was not the site operator at the time of the spill, it made no difference. It created fear in the hearts of the people, a fear that was fanned vigorously by those in power.

Attached to both the people and the oil industry are the politicians. Politicians need people's votes. There were rumors flying that certain politicians would not get voted in if they didn't close the site. Politicians understand and react to threats like that.

Then there is the aspect of returned favors. Eighty-five percent of the state of Alaska's revenue comes from the oil industry. Everyone gossips that elections in Alaska are controlled by the oil industry. The amazing thing about the political control of the state is that it seems to be accepted as part of the normal course of events by Alaskans. Books have been published showing which politicians are controlled by the industry. It's these politicians who get the most money to run their campaigns, so it's these politicians who continue to be voted into office. Once in office, they are free to pay the oil industry back the favors.

Brown's operation at the Sterling site was not perfect. Although he tried hard to run a clean dump site, the effects of the dump's history—and his own inexperience—twice confounded his idealistic aims. He accepts responsibility for two violations during his tenure, but believes, after his vocal posture against illegal dumping, he was cited repeatedly on what he considers trumped-up charges in an attempt to get the site shut down and ruin his credibility.

The first incident occurred in 1981 at the site itself. Brown had two very large and very deep old drilling mud trenches on the south end of the site. He wanted to dry out the old drilling mud, but in order to do that, he first had to dewater the pits. The water was probably five to ten feet deep in those pits, with the mud down under the water.

Brown dug a flat, shallow pad parallel to the upper mud pit as a place to dry any mud he removed from the big old pit. Uphill, a short distance from the drying pad, he excavated a small round pit or pond, perhaps forty or fifty feet in diameter. The water in the big mud pit could not leach into the ground because the drilling mud in the pit effectively sealed off the bottom and sides of the pit. He had to pump the water into a new, clean-bottomed pond, so that it could leach into the ground. On the site he had a D-3 cat dozer, a track-mounted 690 John Deere backhoe, and a three-inch homelite pump. His routine was quite simple. He put the suction hose from the pump into the dirty water in the big mud pond, and put the end of the discharge hose in the new pond. While the pump was running and the new little pond was filling, he ran the backhoe and brought up old drilling mud from the pond bottom and dumped it on the shallow pad. From there, the water would run off the mud, and he would spread it thin with his dozer so it would dry out.

One day he was too busy to go up to work on the mud himself, and sent one of his employees, Ron Hallabe. The little pond into which the water had been pumped should have been bone dry, but a spell of rain had filled it up. The employee came back and asked Brown what he should do. Brown was busy, and told him to do whatever he had to, to get rid of the rainwater.

Ron went back to the site, reversed the pump, putting the suction hose into the little rainwater pond, and laid the discharge hose in the dirt. Then he started the backhoe and began digging mud out of the old pit.

Ron was a good employee, always doing more than was expected of him. He was trying to get rid of the clear rainwater. It was hard to consider it a spill, or an unlawful act of disposal.

However, as this was occurring, one of the Sterling residents flew over in his airplane, saw what was going on, and promptly went back home and called the DEC. The DEC promptly called Mr. Faultless and Perfect at the borough, who tried to call Dave at his office. Brown wasn't there, so Mr. Faultless and Perfect attempted to call several times that afternoon. When Brown arrived back at the shop at around 5 p.m., he returned the engineer's call, but he had left for the day. Betty told Dave that something serious had happened at the dump site, but she didn't know what.

Brown went home around 5:30 p.m. and made his second mistake for the day. He called Mr. Faultless and Perfect at his home to see why he had wanted him so badly.

He answered the phone by saying, "Well, you really fucked up this time and the way your lease is written if you fuck up bad enough with the DEC, we can cancel your lease. We want to do that anyway, and now we've got the grounds to cancel the lease. John Dough with the DEC went to the district attorney in Kenai this afternoon and filed charges against you."

After telling Brown all of this non-stop, Mr. Faultless and Perfect finally had to come up for air and that's when Brown launched an angry rebuttal.

"Listen, you fat S.O.B. I don't know what I've done, but

I don't appreciate being talked to this way. I haven't committed any crime, so there's no reason for anyone to go to the district attorney. However, in the event something serious has happened that I am not aware of, if I go down because of it, then you and your skinny friend at DEC will go down with me because both of you bastards have held your hand out to me with your palm upturned, only I never put anything in it."

Brown hung up and immediately called his attorney and, acting on his advice, had Chemical and Geological Testlab of Alaska fly a man down to Kenai to take lab samples from the puddles that still lay on the edge of the easement surrounding the site. The samples proved to be rainwater, nothing else. The DEC dropped the charges. But Brown had laid on the line just exactly who Mr. Faultless and Perfect and John Dough were, and the incident was never forgotten. For more than a year, Mr. Faultless and Perfect refused to speak with Brown.

The second incident in which Brown may have deserved citing was related to road oiling. This time the incident took place at his shop, also in 1981. At his shop he had a large, round, five-hundred-barrel, or twenty-one-thousand-gallon tank in which he stored oil for road oiling. Tesoro had given him a considerable quantity of oil from its oily ballast water storage tank. Tesoro had tried to drain the water off, so that Brown would get nothing but oil, but somehow there was quite a bit of water in the storage tank, too. Oil and water separate, leaving the oil on top, with the water on the bottom. When Brown's drivers started oiling the road, they noticed the water in the bottom of the tank and wondered what they should do.

While road oiling is essentially illegal now, in 1981 it was legal to oil your own driveway or parking lot to keep the dust down. No permit was required as long as it was your own property. With that in mind, Dave didn't think dumping the water from the bottom of the tank in his own parking lot would create a problem. It was a Saturday, and the DEC office was closed and the DEC employees had unlisted telephone numbers. Brown tried to call to double check, but he had no reason to think it wasn't okay, so he told the driver to open the valve and let the water run out in the driveway until it turned black, then to shut it off and attach his hose and start drawing oil.

Up the street came a young female jogger in short shorts and halter top. She screamed at the driver and demanded to know what he was doing and who told him to do it. Then she said she worked for the DEC and that it was illegal and that Dave Brown was in trouble.

Brown recognized the girl as Jan Murray. She did office work with the DEC and lived with a DEC official. Brown had encountered the two on a previous Sunday, when they, while driving a state-owned vehicle, had stopped one of Brown's vacuum trucks to sample the load. Brown thought that was being a little overly conscientious on a Sunday, but he had nothing to hide.

On the Monday after the oiling incident, Jack Orney, a DEC official, came to Brown's shop and served him with a citation, and Brown openly voiced his opinion.

"First of all, I don't think I've done anything wrong, but, if I have, I'll accept the blame," Brown told him. "But being harassed really bothers me. Jan Murray is a secretary, and the fact that she lives with you doesn't give her any right to tell my driver what to do. If I had done anything wrong, or if she felt as though I had, then her job was to report the incident to her boss and nothing more. You know, if all the DEC employees didn't hide after hours behind their unlisted phone numbers, I could have called and asked first to make sure I was doing the right thing."

Shortly after that incident, Orney was replaced and Jan Murray was transferred to another DEC office. Her original salary was $1,564 per month. Shortly after moving, her salary was raised first to $1,761 per month, and shortly after that to $2,166 per month.

The next man Jan had a relationship with also worked for the DEC. As Brown's harassment by the DEC worsened with the years, Brown couldn't help but wonder if the decisions made by the DEC concerning the Sterling site may have been influenced by his words about Jan Murray.

Although Brown continued to freely express his opinions, the two incidents didn't increase his favor in the eyes of DEC officials. It was basically the beginning of six-and-a-half years of harassment which turned a fifty-four-year-old college graduate with the desire only to hunt and fish and enjoy the good life of Alaska into a man fighting mad over a continual onslaught of false accusations and innuendoes while the true criminal polluters—the oil companies—continued to be pampered and babied while laying waste to the land and water around them.

Up until 1980, Brown had been happily married, solvent, contented, interested in fishing and hunting, trying to run a clean dump, and tell the good news about Alaska to everyone he could. He had never been forced to deal with bureaucrats, politicians, assembly members, and people intent only upon putting Brown out of business. He had never before had to deal with outlaw oil companies who broke the law every day. He watched as the DEC did nothing about it.

He was not accustomed to, nor was he about to, send any money, gifts, or trips in anyone's direction, regardless of how much influence that might or might not have had on his permit, his lease or his cash flow. He believed in treating everyone as an equal, but did not brown-nose anyone. If he bought a beer for anyone, he expected them to buy one back for him. On Dave's one expedition with a DEC employee, a deer-hunting excursion with Howard Keiser, Howard paid more than his share.

But as Dave's harassment continued, so did his angry self-defense, and, because he didn't choose his words carefully and step around the landmines, he began to antagonize a lot of people, and the real issue, the pollution of the Kenai Peninsula, was disguised.

Why the DEC didn't close down the site in 1980, before Dave took it over, when one of its own officers recommended that it be done, is unclear. With the change of ownership, there might have been legal problems with reneging on a lease or permit agreement. The permit stipulations had not been met by Tom Lyan. But then Brown was given a new permit. Maybe there was still the desire to give the oil companies their alibis while continuing to bury waste at drilling sites. Maybe there was the hope Brown could be a source of favors. Maybe the permit was renewed before those hopes were truly crushed. But with the oil companies out to get Dave, it was only a matter of time before some kind of paper violation would be found to legally close Brown's operation of the site.

The way to make it legal was to create a huge amount of time-consuming nit-picking on the site's operation. It would cost Brown a lot of time and money fighting and proving he was operating the site according to the permit, plus, it would create a long, incriminating-looking paper trail, which, to an observer or a judge, would look like permit violations had been continuing for long periods of time.

As Brown helplessly watched, the DEC cut off his cash flow by closing him down. There was no real cause. But, through this action, it was assured he would have permit violations.

A permittee can't meet permit stipulations if the permit doesn't physically accommodate the site. Brown's new permit didn't. The site was limited in size to eleven acres. It was rectangular, surrounded by a chain-link fence and two locked gates. The limitation in size meant capacity was limited. The new permit stated that new ponds and dikes had to be sloped no steeper than three to one, which meant that for every foot down, the bottom edge had to be three feet out. The new permit also stated that drilling mud had to be mixed with dry dirt in lifts or layers not to exceed six inches and be capable of supporting heavy equipment.

There was not room enough in the little dump site for three to one slopes and the DEC had to know it. This would not only severely limit the site's capacity, it would close it overnight. To accommodate the permit, Dave dug four new pits, excavating into the earth instead of building on top, so no dikes as such were necessary. State solid waste management regulations did not specify three to one slopes, nor did state statutes, but instead say that drill mud disposed of on land must be confined in a pit, trench, or diked area that prevents erosion or mass flow of the mud. Dave figured using pits instead of ponds and dikes would keep him safely within the law, especially since the reason for the stipulation—to prevent the erosion and overflow of the mud—would be taken care of with the pits.

Brown talked over his method with Jim Kross, of Environmental Services LTD., who later covered it in the site's plan of operation, which was acceptable to Bob Martin, special assistant to the DEC commissioner.

But Dave also had problems accommodating the six-inch lifts. Those regulations were so designed for sanitary landfills to prevent litter from blowing around, and to allow heavy equipment on it. Drilling mud had the consistency of extremely wet cement, and it would be absolutely impossible for it to support heavy equipment unless it was a mixture of ninety-five percent dirt and five percent mud. Both of those permit stipulations were ridiculous. Dave just overlooked them at the time as bureaucratic garbage that couldn't be taken seriously because they insulted the intelligence of any person familiar with landfills or drilling mud. Later, he felt they were probably inserted as potential legal excuses for closing down the site anytime DEC wanted to.

A DEC environmental engineer was on hand when Dave dug his new pits. He approved verbally and gave suggestions. After the new pits went in and were lacking the three to one slopes, DEC sent Alan Boggs from Juneau up to Kenai Peninsula twice with specific instructions to see if there wasn't sufficient fault with the operation of the site to be able to close it down. Each time Boggs gave the site a clean bill of health. In spite of this, the DEC cited Brown numerous times for not having three-to-one slopes, or operating with six-inch lifts. The violations weren't different ones, just two violations impossible to correct. They were made over and over again. The state of Alaska now had the paper trail it needed not only to shut Dave Brown down, but to discredit him and his oil company accusations.

There was also other regular harassment which, Dave felt, could either be considered purposeful, or just bureaucratic incompetence. For example, Dave was required to regularly furnish certain records to DEC. Once every month he had to supply the Soldotna office with copies of his barrel dumping logs. Every three months he took water samples from the monitoring wells, and sent them by air to Anchorage for testing by Chemical and Geological Laboratories of Alaska. Copies of those test results had to be sent to the Anchorage DEC office.

The barrel dumping logs were always promptly sent to the Soldotna DEC office, but on numerous occasions John Dough called and demanded them again. He claimed, "I don't have them." Maybe it was harassment. Maybe it was incompetence. Maybe he thought he could catch Brown with two sets of logs and "get him."

Brown turned in monitoring well reports quarterly, although not always on the same date. Sometimes the first day of a quarter fell on a weekend, and sometimes the test lab was swamped with other tests, but the results were turned in within a few days of the beginning of the quarter.

Out of the blue, Bob Martin issued an ultimatum. "A quarter always starts on the first day of the month and you have only five days after that in which to get your test results to me."

Brown thought the ultimatum was pretty ridiculous, and, as is his way, let people know it, but complied with it. Plus, he figured five days after the first of the month would equal the sixth of the month.

At 3 p.m. on July 6, when the site was extraordinarily busy, John Dough met Brown's vacuum trucks at the gate of the Sterling site and sent them back to Union Chemical because of a phone call made by Martin to Dough that morning.

"You're shut down," Dough told Brown. "Your test results weren't turned in on time."

"You S.O.B." Brown muttered half under his breath, then responded to Dough. "Those results were hand delivered to Mr. Martin at 11:00 a.m. today. Not only that, I have until the end of business today to comply with Martin's ridiculous decision. That's five days after the first of the month."

Dough stood his ground, at least for forty-five minutes.

At 3 p.m., Martin had had his hands on the test results for four hours. A simple phone call back to Dough would have prevented the entire scene. Instead, Ernest Mueller, the former DEC commissioner and business associate of Environmental Services, was in Anchorage that day, and a quick phone call sent him to the DEC Anchorage office where he bawled Martin out. Forty-five minutes later the site's gates were reopened.

Yet, the episode was one more item in the paper trail to discredit Brown and the site [See Appendix F].

Another instance involved Brown's shop in Kenai. In the parking lot outside the shop was a low spot in the driveway. Every time it rained or the snow melted, there was a puddle. It would have taken many loads of gravel to eliminate that low spot, and Brown was doing so as he could afford the gravel. One day, John Dough and the other DEC officer came from the Soldotna DEC office to take samples of the mud puddle, thinking Brown had dumped some hazardous waste in the driveway. They found only water. While they were doing these kinds of things, they were complaining about how understaffed they were, so they could not adequately regulate the many drill sites and refineries.

The nit-picking was harassing and irritating, and ultimately cost time and money. But the real problems the DEC caused Brown involved cash flow. Soon, the skirmishes had erupted into an all-out war between Brown and the DEC. Brown was armed with information on the oil companies. DEC was armed with the ability to change its mind, costing any permittee hundreds of thousands of dollars, with the ability to slap on fines when it wanted, and it was armed with Brown's own stubborn, crusty personality, which DEC officials used skillfully to discredit anything and everything he said.

The battleground was no longer just the paper trail of supposed permit violations. Instead it became a public forum. While DEC was nit-picking Brown's Sterling site, officials were denying any knowledge of illegal dumping by the oil companies, in spite of the fact Brown had been exposing Unocal's illegal dumping in the Kenai Gas Fields, along with other oil company dumping, since 1981. Howard Keiser started to investigate, but then, for reasons never satisfactorily explained, Keiser retired in 1982 and all investigation stopped.

However, Keiser did write a letter to the DEC later, and a copy was published under "Letters to the Editor" in *The Peninsula Clarion.* It states:

"I wish to object to the issuance of a disposal permit to Union Oil Co. of California for disposal of gas well cuttings and solids off of Kalifonsky Beach Road.

"My reasons for objection to the issuance of this permit are as follows:

"1. There have already been hundreds of thousands of barrels of this material buried in this area in wetlands, with no permit.

"2. The area in question is a spring resting area for snow geese, tule geese, Canada geese and thousands of other waterfowl.

"3. The area in question is the only nesting area on the Kenai Peninsula of the small colony of lesser Canada geese, other waterfowl and other water-oriented bird life.

"4. During spring breakup there is now always a visible oily sheen in the areas of the present buried drilling mud and cuttings. This can be detrimental to the eggs of the nesting waterfowl and other nesting birds. It has been shown through research that one drop of oil can kill an embryo of an egg.

"5. The area in question is in wetlands. Even where there is no standing water, the water table is only 3–6 feet below the surface of the ground. The quality of water in this area now is poor. Burying this material within the water table would jeopardize the safety of the water in this area.

"6. Run off of the area in question is into Cook Inlet. If there is any leachate in future years, it will go into Cook Inlet, which could be harmful to the flora and fauna of this body of water.

"7. The basic reason for Union Oil requesting this permit is that it will be less expensive to dispose of these materials on site. If this is permitted, in the near future there will be no nesting waterfowl or other birds. People will not be able to utilize water from their wells, which is common throughout the Lower Forty-eight due to disposal of materials. If this is permitted, it will set a precedent. Every oil company, chemical company and refiner will have their own dump site throughout Alaska. Controlling and monitoring so many sites would be impossible."

By 1984, Brown had come to the conclusion that the waste situation in Alaska had gotten completely out of hand. He was convinced that not only Union Oil, but the oil industry in general all over Alaska and particularly on the Kenai Peninsula, was guilty of reckless and dangerous environmental practices and unlawful disposal. Brown had been forced by the DEC and by the Kenai Peninsula Borough to invest more than $400,000 at the Sterling Special Waste Site, mostly to meet standards which didn't really have anything to do with the potential for pollution of the groundwater [See Appendix G]. Yet the oil industry got a free ride, dumping where they wanted.

Finally, in October, 1984, Dave couldn't take the garbage being heaved at him any more. He spoke to Betty.

"I want to go public, but you know that if I do this, it could ruin us financially," Dave warned.

As ever, Betty was supportive. "But what if you don't? Could you live with yourself? The DEC isn't doing anything to protect this state. Someone has to."

Her support was all Dave needed. He decided to go to war. He hired an attorney and went public with his information, buying newspaper ads in *The Peninsula Clarion*, telling the people the truth about the oil industry and telling the people the truth about the Alaska Department of Environmental Conservation [See Appendix H].

His ads specifically exposed the illegal waste dumping in both the Cannery Loop Units in the city limits of Kenai, and the Kenai Gas Fields, both operated by Unocal, then Union Oil. One ad, in part, stated:

"My first experience with Union Oil Company was in May 1981, when, at Joe Callendar's request, I personally measured the amount of drilling mud located at their Cannery Loop #2 reserve pit. I wrote Mr. Callendar on May 22, 1981 and advised him that his pit contained approximately fifty thousand barrels of drilling mud, and I gave him my price for removing all the mud from the reserve pit, and transporting it to the Sterling Special Waste Site for disposal. He then contracted with me based on the information I supplied to him to remove all the mud from the Cannery Loop #2 reserve pit. This contract was issued using an hourly rate instead of a flat rate.

"The dimensions of the Cannery Loop #2 reserve pit when I measured it in May 1981 were four hundred feet long by eighty feet wide, and it contained approximately eight foot of mud therein. A few days later, Mr. Callendar contacted me by phone and advised me to start removing the mud from the pit. When I arrived, the pit that two days earlier was 400 feet long had suddenly shrunk to a mere two hundred feet. Union Oil Company contractors had backfilled approximately fifty percent of the pit before we started pumping, and they continued backfilling as we pumped what was left.

"On 6/10/81, I again wrote Mr. Callendar and advised him of the total amount of mud removed from Cannery Loop #2 as well as the costs to remove it. We disposed of 15,823 barrels of drilling mud from Union Oil's Cannery Loop #2 reserve pit. The other approximately 34,000 barrels were buried by Union Oil Company, on location, and against the DEC's regulations, with no permit. This is a residential area on Beaver Loop Road in Kenai, which has a high water table, and shallow drinking water wells.

"After Cannery Loop #2, Union Oil Co. moved closer to Kenai and drilled Cannery Loop #3. This well is located on the Bridge Access Road across from the Chevron Oil Bulk Plant. Out of this well we disposed of approximately 8,000 barrels of used drilling mud, the rest to remain behind a chain-link fence to slowly leach into the local water table."

In that same ad, Dave continued listing the different

wells drilled by Union Oil, moving them into the Kenai Gas Fields, where the company received fifteen permits to drill, mostly in swampland. Dave received eleven thousand barrels from the drilling in that gas field, which means he received from Union Oil a total of 34,823 barrels of mud, less than the amount one would expect from one well. The ad continued:

"Union Oil Company has, in the past three and a half years, disposed of, with no permit, approximately 500,000 barrels of used drilling muds. This is 21,000,000 gallons. Where is it all going? Certainly not to the Sterling Special Waste Site where it is supposed to be going. They are now asking the Alaska Department of Environmental Conservation for an after-the-fact permit to leave there what was illegal in the first place. The area they are requesting a permit to dispose in has a high water table and poor soil conditions for this type of disposal. No permit could be issued by any individual or agency without making a one hundred eighty degree change in the existing regulations.

"Union Oil Company was advised by Howard Keiser (who was with the DEC at that time), by letter dated 5/4/79 and 12/7/81 that they were in violation of DEC regulations if they did not dispose of their drilling mud at the Sterling Special Waste Site. He started an investigation, but retired in 1982. To date, there is no indication that DEC ever followed up on that investigation.

"In early July 1984, Howard Keiser and I walked through the Union Oil Gas Field marshes with hip boots, cameras and binoculars. These marshes are among the most beautiful in the state, and they have always been a nesting and resting ground for many species of water fowl. Last year the bird numbers had decreased noticeably. This year there doesn't appear to be any birds in the area at all. We saw one duck, a lone pintail, and he was only passing through. There were no geese at all. There were no terns. There were no sea gulls. There were four sandhill cranes, and four or five sandpipers. The cranes had no eggs and no babies, because they made none of the threatening gestures that cranes make under these circumstances. They are smarter than we are, so when Union Oil Company fouled up their environment, they simply left.

"There is an intermittent oily sheen started at Union Oil's last mud pit and extending out about one mile into this beautiful marsh. There is drilling mud up three or four feet high on the tree trunks near Union Oil's last mud pit, indicating that the mud pit has been allowed to overflow.

"It appears that drilling mud is not the only thing being covered up. We as taxpayers are paying the DEC to protect our environment, while for some reason, which they don't care to discuss, they are deliberately looking the other way, and allowing our environment to become destroyed."

Dave ended that ad with a plea for people to submit objections to issuing the after-the-fact permits, either in writing, or by attending a public hearing October 29th. He concluded, "Come to the meeting on October 29 in Kenai, stand up and be counted!"

The ad had the expected ill effects and the hoped-for good effect. Dave's phone started ringing, and the first caller was from Union Oil, cancelling a job. He expected that. But then came the positive results. People began calling, thanking him for bringing a serious problem out into the open. He even received a couple of phone calls from Union Oil employees saying thanks, that it was about time someone blew the whistle on Union. But the words, "Don't mention my name . . . I need my job . . ." always ended the conversation.

Betty received a phone call from a woman who lived near Cannery Loop #2.

"My drinking water smells like diesel oil, and now I know why. But I can't complain to the DEC or the oil company. My son works for them."

Recognition and notoriety began to follow Dave. Complete strangers called him and stopped him on the street and congratulated him.

8

THE FIGHT IS ON

Dave wasn't content with a one-time stirring of the chemical pot. He knew he would have to continue, especially to get people out to the public hearing. He ran more ads. Instead of attacking the oil company, the new ads attacked the DEC. One, in part, read:

"For the last four years as the operator of the Sterling Special Waste Site, I have been deliberately and ridiculously harassed by the DEC and I have been subjected by them to what I call, Selective Enforcement of the Law, i.e.: You do this Dave Brown, but no one else has to.

"This in itself is a violation of my Constitutional rights under the 14th amendment to the United States Constitution, which says, no state shall deny to any person within its jurisdiction the equal protection of the laws . . .

"The reasons for which I am being harassed and selectively enforced are very simple.

"1.) I don't treat the D.E.C. with the respect which they feel they are entitled to. My reason: So far I have met no one employed by the D.E.C. whom I felt was deserving my respect.

"2.) I know too much. As the operator of the Sterling Special Waste Site, with the records I keep, I know who is obeying the rules, and I know who is violating them. When I ask the D.E.C. why these violations are allowed they refuse to answer me.

"3.) I am very outspoken. I tell it exactly like it is and I let the chips fall where they will.

"My evidence indicates that certain Oil interests may have made it worth while for certain D.E.C. people to 'close their eyes,' and let the law be violated in order to save hundreds of thousands, even millions of dollars for these oil interest . . ."

Dave then offered a positive solution.

"We need someone close by to watch the D.E.C., because as has already been proven, they are inept, incapable, and unconcerned, for whatever reason.

"We have federal regulations, state regulations, borough regulations, and in some cases municipal regulations. No single agency has the right to allow any other agency's regulations to be violated.

"Since the Kenai Peninsula Borough boundaries are far reaching, and since they have offices in Soldotna with an Assembly and a Mayor that are always accessible, then I say let's let the borough get involved. Let's not be restricted to dealing with the D.E.C. people who are always either uncooperative or inaccessible.

"Recently the Borough has decided to establish a Solid Waste Commission, a long overdue good idea. Let's consider having that commission involved in solid waste and liquid waste. Let's give them the power to investigate nasty situations. Then let's give them a free rein to go to the D.E.C. with the problem, and then to go over the D.E.C.'s head to the proper higher authority if the D.E.C. refuses to act immediately.

"I am very favorably impressed with what I see on the Kenai Peninsula Borough Assembly, people helping people for the good of all of us.

"I am not impressed favorably at all with the D.E.C. an never have been. We need a housecleaning in that department."

Again, Dave made his plea for people to come to the October 29th public hearing, then ran his version of Joyce Kilmer's poem:

"I think that I shall never see
Drilling mud as lovely as a tree
A tree that may in summer wear
A nest of bentonite in her hair
A tree whose hungry mouth is pressed
Against the muds fast rising crest
Who lifts her leafy arms to pray
Lord get me out of this mud today
Upon whose bosom snow has lain
Who ultimately dies in vain
Poems are made by fools like me
But only Union's Mud can kill a tree!"

Another ad, which ran October 29th, was patterned like an open letter to a Commissioner of the DEC, and to a drilling superintendent for Union Oil. The first sentence states, "I am a little tired of the two of you insulting the intelligence of the People of the State of Alaska by breaking the law and bending the rules to suit your own purposes."

Dave's homespun but caustic wit was evident when he, as if speaking to children, recited the ordinances the commissioner was supposed to uphold, and the oil company official was supposed to at least know, as a superintendent. The regulations, Brown stated, say, "A permit is not required for a single family or duplex residence or farm where solid waste is generated and disposed of on the premises. I haven't seen any cows, pigs, or chickens in the Union Oil Gas Fields, so I don't think the Union Gas Field meets these criteria, but they have been disposing of mud and not storing for years regardless of what stories they tell to the contrary."

In addition to buying the newspaper ads, Brown and his attorney, William Bixby, circulated a thousand copies of the October 23, 1984, issue of the *Wall Street Journal* regarding the disposal of drilling mud in Louisiana, and the resulting contamination of the ground water.

The ads, the distribution of the articles, and coverage by the local newspaper had the desired effect.

When Dave, his attorney Bill Bixby, along with about one-hundred-twenty-five fired-up residents jammed into the Kenai City Hall, ready to burn in effigy Union Oil and the DEC, they arrived just in time to hear a local engineer describe in great detail how a bentonite cap over the site would preclude the need for any kind of liners in the pits, because the drilling mud was mostly bentonite, anyway. He came armed with colored drawings of what the site would look like, and how it was supposed to work.

Dave stood up and said loud and clear, 'We don't buy it." A dark-haired woman in the rear continued, "We're adamantly opposed to all disposal of drilling mud in the city."

Then, toward the end of the meeting, Dave rose again. The crowd hushed.

"I have evidence," he said forcefully, "that there are DEC officials on the take from the oil companies. It's right here on this tape recorder."

Bixby turned on the recorder. Unfortunately, the quality of the tape was poor. People had a hard time understanding exactly what was taking place and an even harder time understanding who was speaking.

Despite this, the result of the hearing was that more people like Leigh Smithers finally began to realize what was being dumped in their own backyards, and that DEC officials were probably overworked or inept or worse.

The hearing was followed by a barrage of newspaper articles, editorials and cartoons regarding Union Oil and drilling mud. The Alaska Fish and Game Habitat Division got involved. They stated that they strongly opposed the permit request. The city of Kenai, through its mayor, Tom Wagoner, asked the state to deny Union Oil's permit requests. The Kenai

Native Association joined the protest, and probably more important than any of it, the landowners, the Salamatof Native Association, Cook Inlet Region, Incorporated, which is a Native corporation, the Kenai Peninsula Borough, and two private landowners, Doris M. Lashley and Leonard A. Keener vocally opposed the permit. Permit applications require the consent of the landowners, and both Lashley and Keener said they weren't even approached by Union Oil. In an article in *The Peninsula Clarion,* Keener stated, "They haven't approached me and I wouldn't sign it (a letter of permission) if they did. I'd feel bad if they polluted the water."

Lashley commented, "I am very unhappy with Union Oil. I would never, never grant permission." Cook Inlet Region was equally adamant about never granting permission.

Even the Federal Bureau of Investigation displayed some interest in the case, although nothing ever came of it.

After Dave's ads appeared, Union Oil finally took some action. A frenzied, twenty-four-hour-per-day rush began in the Union Oil Gas Field. Some vacuum trucks belonging to Dave's competitors were moving something from one point to another, but definitely not taking it to the only legal drilling mud final resting place on the Kenai Peninsula. In fact, one competitor told Dave that he was grateful for the newspaper ads, because as a result of those ads he got a lot of work that he wouldn't have gotten otherwise.

In the midst of the public racket, Dave got a long distance phone call from Mike Pelch in Eugene, Oregon. Mike was the owner of the land on Beaver Loop Road upon which Union Oil had drilled the well known as Cannery Loop #2. Union had leased the land from Mr. Pelch, but had let the lease expire.

Pelch's deep voice crackled, "Dave, your newspaper campaign has been causing me some problems. I've invested a considerable amount of money in my land for engineering in order to gain borough approval for subdividing. Now, with that story out about the dumped drilling muds, I wouldn't be able to give my lots away."

Dave mused. "You know I didn't have anything to do

with it," he answered, "but, hell, people who live here all the time have a right to hear the truth." Pelch was an absentee land owner who came to Kenai in the summers.

Pelch agreed and said, "I didn't grant permission to Union Oil to dispose of any drilling mud on my land, and I was not aware that they had done this."

Pelch queried, "Look, Dave, How about you digging up Cannery Loop #2 and exposing the drilling mud. If the mud's indeed there, I'll file suit myself against Union Oil."

"I'll do it, Mike," Dave answered forcefully, "and I'll do it right."

On January 21, 1985, after calling *The Peninsula Clarion, The Anchorage Times,* The Alaska Department of Environmental Conservation and the United States Army Corps of Engineers, Dave went into Cannery Loop #2 with a D-8 cat and backhoe. After ripping the frozen ground with the cat, he slowly moved the backhoe into place, he dug a hole about thirty feet long and fifteen feet deep.

"Hell, here it is," he yelled to the reporters.

The drilling mud was just where Dave had said it would be. *The Peninsula Clarion* article, which was published the next day, explained just what Dave had done, and what was found. The article went on:

"Union spokesman Mark Eversole discounted the find, however, and said the company had a permit from the Alaska Corps of Engineers allowing disposal of drilling mud solids at the site.

" 'We've never denied that there was drilling mud there,' Eversole said. "That was spelled out in the application. We stated there would be a pit, that we would pump off the pumpable liquids and bury the solids.'

"Pat Richardson, assistant public information officer for the corps denied any knowledge of a Union permit. According to Richardson, the corps grants permits only for clean fill material.

" 'We don't regulate drilling mud,' she said. 'Drilling mud would have things in it that we wouldn't allow under our permit.'

"Although the State Department of Environmental Conservation is considering a Union permit application to dispose of drilling mud at seven sites, Cannery Loop Unit Two is not part of the proposal. The Alaska Oil and Gas Conservation Commission has approved abandonment of the site in 1982.

"C.V. Chatterton, chairman of the commission, said a site is approved for abandonment when the surface is clear and free of all trash and debris. Chatterton said the AOGCC is not responsible for what might be buried underneath abandoned drill sites.

" 'I don't know what would be,' he said. 'That's a good question. As far as this agency is concerned the site was cleaned up. We don't worry about waste disposal. That's the turf of another state agency.'

"Local D.E.C. officer Bob Cannone said he was not certain which agency bears responsibility for making sure the site was cleaned. He said he would be interested in seeing proof that drilling mud was found at the site.

" 'Whoever issues the permit is responsible,' he said. 'It would be their (AOGCC) responsibility to police the area.' "

The article continued, stating the DEC officials are developing guidelines for drilling muds disposal, which were to be unveiled in April, despite the fact the state's own solid waste regulations already applied to drilling muds.

However, Dave Brown's crusty personality didn't win him many accolades. Some members of the Anchorage press made thinly veiled allegations questioning his motive and legal right to dig up the pit. True to his fiery nature, Dave took out another newspaper ad to answer their charges. It stated, in part:

"To begin with, Michael Pelch the landowner in fact asked me to dig up Cannery Loop No. 2. He also ordered the D.E.C. and Union Oil to remove immediately any and all mud that was found.

"D.E.C. says they wanted to know the truth about Cannery Loop No. 2. Nonsense. I told D.E.C. the truth about Cannery Loop No. 2 four years ago. Howard Keiser, who worked seventeen years with the D.E.C. told Union Oil to move the mud. Union Oil ignored Howard. Then Howard retired and the issue died."

Not stopping there, Dave also wanted to rip a hole in DEC's "April unveiling" of drilling mud guidelines theory as an excuse for ignoring drilling muds since the 1974 solid waste regulations. He ran another folksy ad. This one said:

"Yes, Virginia, there is a Santa Claus here in Alaska. His toy shop is up in Anchorage at the Alaska Department of Environmental Coverup.

"In Alaska, Santa is not called Santa Claus, not Kris Kringle, not St. Nick. Instead, he is called St. Martin.

"St. Martin has two little elves to help him out, Lamarr O. and Flintstone.

"It is St. Martin's job to give out toys to all the little oil companies that have been good.

"Right now, Virginia, Pollunion Oil is trying to get St. Martin to give them a drilling mud disposal permit.

"This, Virginia, is a very difficult task for St. Martin, because this is what is known as an 'after the fact permit.'

"It's especially difficult now because the Department of Environmental Coverup enacted their permit regulations in 1973, and this is 1985. Pollunion Oil has drilled about 20 wells since 1973, with no disposal permits for any of them from the Department of Environmental Coverup. To add to the difficulty, Virginia, it's not even Christmas. It's March and Pollunion is in trouble. So is Coverup. This will take a lot of 'Our Fathers' or 'Hail Marys' and a lot of political maneuvering.

"Have faith, though, Virginia, in St. Martin, and in his two elves Lamarr O. and Flintstone. They have a way of getting things done for good little oil companies. Also, Virginia, St. Martin loves Pollunion, because the Lord loves a sinner and St. Martin himself is a Deeply Religious Man.

"Just think, Virginia, by April 15, thanks to St. Martin, Lamarr O. and Flintstone, Pollunion Oil will probably have their soul cleansed and their sins forgiven for the past 12 years, without even going to confession. They will also probably have a shiny new permit.

"This is how the Golden Rule operates, Virginia. He Who Has The Gold, Makes The Rules.

"You were always taught, Virginia, that Faith Moves Mountains, and it does, but sometimes Money Moves Them Quicker.

"Just think, Virginia, Franklin D. Roosevelt changed the date of Thanksgiving, but St. Martin may change the date of Christmas from December 25 to April 15, at least for Pollunion Oil. Twelve years of Christmas for Pollunion, all at once, and in Juneau they can trade some of the egg they are wearing on their faces, for drilling mud. Perhaps they'll even have some of that drilling mud on the menu at the Baranof—Bentonite a la St. Martin!

"Let's you and I pray for them, Virginia, for they know not what they do. Let's also pray, Virginia, that this new permit gives peace of mind to St. Martin, to Governor Sheffield, to the Alaska Department of Environmental Coverup and to Pollunion Oil.

"And oh, yes, Virginia, let's you and I send a copy of Grimms Fairy Tales to Pollunion Oil, to help them with future press releases, just in case St. Martin can't do the job. One Hour Martinizing doesn't always work.

"One more comment, Virginia, before I sign off.

"St. Martin was appointed some time ago as Special Assistant to Commissioner Today, (for here today, gone tomorrow).

"Commissioner Today has since been reappointed as Special Science Fiction Advisor to the Governor.

"Now, in the face of declining oil revenue and impending budget cuts, Governor Sheffield can no longer afford his Science Fiction Advisor, so here Today will be gone tomorrow.

"Virginia, since the budget cuts are necessary, and since the Commissioner who St. Martin was supposed to assist is no longer here, do you really think we can afford to keep St. Martin here to assist a man who is long gone and beyond lending assistance to?

Dave Brown
Alaska Environmental Industries
Keeping Alaska Clean
For Your Kids & Mine

"P.S. My apologies to Governor Sheffield and to our new D.E.C. Commissioner Bill Ross if this offends either of you. I wrote this in January, before either of you publicly stated that you were going to help us with our P.P.P. (Peninsula Pollution Problem).

"I believe you were both sincere, and I vowed I'd say no more.

"Now however, it's obvious that the Sterling Special Waste Site test results are being held up until April to coincide with D.E.C.'s new drilling mud guidelines.

"It is also obvious that if D.E.C. couldn't do the job before, with a reasonable budget, that they certainly can't do any job at all with substantial budget cuts.

"Last of all, after reading the recent 'Lucky & Lamoreaux' Cannery Loop 2 comments in the Clarion, it's perfectly obvious that the policy of the D.E.C. and the State of Alaska shall be to 'PROTECT FOREVER' the Oil Industry from the people of the State of Alaska."

Finally, on May 1, 1985, the DEC issued the following press release:

"The State of Alaska Environmental Conservation Commissioner Bill Ross and Attorney General Norman Gorsuch today announced initiation of a civil lawsuit against Union Oil Co. and other lease partners for the illegal disposal of oil and gas drilling muds on the Kenai Peninsula. The charges filed in state court concerned a drilling muds disposal site, "Cannery Loop No. 2," near Kenai. Union Oil Co. was the unit operator, in partnership with several other companies, in the drilling venture at this site.

"Attorney General Gorsuch and Commissioner Ross said, "The State filed this suit after an extensive investigation confirmed that Union Oil, the site operator, improperly disposed of oil and gas drilling muds without first obtaining the necessary DEC solid waste management permit.

"Drilling muds and cuttings are the solid and liquid waste products of gas and oil well drilling operations. Muds and cuttings contain chemicals and other elements that facilitate the drilling process but which can cause serious pollution when improperly disposed."

Dave felt fighting mad over that release. He had spent four years telling the state about Cannery Loop #2, and in all that time they hadn't given a damn. As for an "extensive investigation" it was Brown's work, digging it up. He had spent $3,205 of his own money to expose Union Oil.

But the part of the release that truly sickened Dave was the filing of the civil suit.

"Crap," he muttered to Betty, "A civil suit will be tied up in court for years. It should have been a criminal suit. Pollution is a crime. It is recognized and prosecuted as a crime in every state except Alaska, where the state deliberately looks away and allows it to happen on a regular basis. In Alaska, the statute of limitations expires, so that these criminal acts may no longer be tried criminally, but may be tried as a civil suit, if handled at all."

Betty countered, "Maybe it will be different this time."

But what Dave expected did happen. After filing civil suit against Union Oil for Cannery Loop #2, and denying the permits for the rest of the pits, the DEC sat back. Action had been taken, so to speak. But the drilling muds remained where they had been placed, as late as the time of this book's publication. The reason? "It's in litigation. It's been appealed to the Superior Court," the DEC officials inform reporters or interested citizens year after year. "We're negotiating with them," or "We're working with them," they say over and over. There's been no clean up, and no fine has ever been paid [See Appendix O].

Of course, Brown knew that to most people he was just an irritating gadfly whom the DEC wanted to ignore. He also knew he could and would be ignored by authorities as long as the public ignored him. But he was aware if the public began to take notice that, just as Brown was telling them, their treasured Alaskan lifestyle and their wildlife and fish were being threatened, then more serious action against him would probably be taken.

And the public was starting to take notice that Dave Brown might be a damned, loud-mouthed nuisance, but a lot of what he was saying was true. And public ire was raised.

DEC officials began to take Brown more seriously. Actually their attempts to discredit him had begun earlier than Brown had originally realized. While digging through files in later years, Dave ran across a memo from one executive in DEC's Anchorage office to another in Environmental Quality. The memo is dated August 28, 1984, several months before his newspaper ads, but several years after Dave began loudly complaining to the DEC about illegal disposal of muds.

The memo states:

"Mr. Dave Brown, who operates the Sterling Special Waste Site, is evidently concerned that illegal dumping of drill muds is occurring. We have repeatedly told Mr. Brown that we will investigate any incident of illegal dumping of which he is aware. My understanding is that he has hired an investigator to document his contention, however, as of this date, I've seen no information from him.

"Mr. Brown also objects to our allowing other permits for disposing of drilling muds, contending that the safeguards provided at Sterling are not present at other permitted sites. I disagree with him on this. We have presently permitted the disposal of drilling muds in two injection wells, two reserve pits for exploratory wells, and have issued a compliance order pending permit review for disposal at several locations with the Union Oil operating area. These disposal activities differ from Sterling in several regards:

"1) They all handle drilling wastes from a single source rather than accept wastes from a wide variety of sources.

"2) Each of these sites is under the control of the entity producing the waste, whereas Sterling is under the control of someone who has no detailed knowledge of the wastes being disposed of.

"3) The reserve pits used for disposal are fully lined and sealed with an impermeable cap when completed but Sterling wastes are deposited in unlined pits.

"4) Reserve pit disposal is a one time disposal operation as opposed to a continuing operation.

"5) Disposal in reserve pits is at the site of waste production and this eliminates the need for additional handling and transport for disposal elsewhere.

"The Sterling site has always involved extensive monitoring and inspection. We have no indication of a contamination problem at this time, however, Mr. Brown is only marginally in compliance with operating requirements. Additionally, it appears to me that the site has very limited additional capacity. Local residents have, of course, been actively pressing for the site's closure.

"Recent inspections have revealed a number of concerns regarding the manner in which the site is being operated. We sent Mr. Brown a letter itemizing these perceived violations (attached). Our letter stimulated a meeting between our staff and Mr. Jim Kross, who serves as Mr. Brown's engineer. At this meeting Mr. Kross asserted that the site really is in compliance with permit requirements and that our concerns are the result of misunderstandings. In my opinion, there is some likelihood that misunderstandings have occurred in some respects and that in others there are needed corrections. There is also a need to consider revising our permit to clarify our expectations. We are continuing to work with Mr. Kross on those items that need correction, and to revise the wording of our permit to achieve greater clarity."

Brown had a real hard time swallowing that memo. He yelled at Betty and everyone else who came near, "It's pocked with damned lies and half-truths. DEC officials never told me that they would investigate any incident of illegal dumping which I reported. In fact, I repeatedly told them about such incidents but they repeatedly failed to investigate."

DEC officials even told reporters that Dave Brown was just blowing off hot air, that he had never been specific about any real illegal dumping.

"And that's another filthy lie," Dave protested. "In fact, in the second paragraph, the author of that crummy memo refers to issuing two permits for injection wells, two reserve pit permits for exploratory wells, and he refers to a Union Oil compliance order," Dave fussed to his wife. "Well, a sorry excuse for a job undone. A total of ninety-nine wells have been drilled since 1973 within the boundaries of the Kenai Peninsula Burough—thirty-two on land owned by the state, and thirty in the Kenai National Wildlife Refuge. Eight permits were issued by the DEC for drilling mud disposal. Four of these were across Cook Inlet at Beluga River and Trading Bay. Ten permits were issued by DEC for waste water injection. Drilling mud disposal permits and waste water injection permits overlap—both are often needed for one well, since injection takes care of the liquids and burial takes care of the solids.

"Then the memo states—as if it is a point in favor of on-site disposal—that 'each of these sites is under the control of the entity producing the waste.' By this official's own statistics, in eighty-nine of the ninety-nine instances the 'entity' buried the waste illegally, on site.

"At the time the memo was written, one pit on the Kenai Peninsula was lined and capped—ARCO's pit at Wolf Lake #2. One pit was capped, but not lined. That was ARCO's Funny River site.

"Item number five," Dave went on, using Betty as his sounding board, "displays either the author's incompetence or his removal from earth. For years, garbage and septic waste have been transported for disposal elsewhere. In the state of California drilling mud is commonly hauled at least fifty miles for disposal."

"Sometimes much further," Betty agreed,

"Why should Alaska be any different?" Dave asked plaintively. "Because it's bigger and has fewer noisy people?"

Dave was on his crusade. "But what is mind-boggling is the tone of the damned memo," he muttered. "That on-site disposal is better than hauling the waste to one, well-monitored location is a lot of hogwash."

It was for this purpose—to control, contain and keep track of oil field waste—that the Sterling Special Waste Site was established just nine years before.

While Brown was buzzing around the peninsula, stirring up oil hornets, the DEC was doing a lot of buzzing itself, trying to do whatever it could to discredit Brown, attacking his personality and his business. Yet Brown wasn't the only person on the peninsula to "go public" with his problem. And he wasn't the only one with a problem and with gripes about the DEC's inaction.

A group of homeowners living near a Unocal-owned gravel pit adjacent to the Kenei Gas Fields had been complaining since 1981 about Unocal's illegal dumping of industrial garbage. The complaints yielded no action once again until the press, and therefore the public, got wind of the atrocity and actually caught the violators "red-handed" in the midst of a midnight dumping [See Appendix P for a series of articles detailing the problems of those homeowners]. Still, no real action was taken, so those residents added their names to the list of lawsuits filed against Union Oil. The list included:

1. Alaska Environmental Industries Inc/Mar Enterprises Inc./Dave and Betty Brown vs. fifty-five named defendants including Union Oil.
2. Mike Pelch vs. Union Oil.
3. Dickson, Englemann and Torwick vs. Union Oil.
4. Joe and Billie Hardy vs. Union Oil.
5. Jessica Black and Bob Sizemore vs. Union Oil.
6. State of Alaska vs. Union Oil.

There were the West Poppy Lane homeowners whom the DEC tried to discredit as alarmists—as scaremongers easily distressed over nothing. And there were others with even more serious problems, who received the same treatment—or lack of it—from the DEC.

9

ON THE PENINSULA

One of the people who read Dave Brown's ads and was at the assembly meeting to watch Dave's battle with the authorities was Leigh Smithen, another resident of the Kenai Peninsula. But to her, Dave Brown was just another well-heeled dreamer spouting idealistic slogans. Still, she admired the strength of will which kept him fighting, but for her living with toxic waste was an everyday, nightmarish reality.

Leigh Smithen kicked at a piece of moss lying on the trail as she considered her next move in this tangled web of underground waste streams into which she had been inadvertently woven. Alaska had not really been too good to her, despite her repeated attempts to live in this land which tantalized people by hanging its beauty in front of them.

Her first arrival June one in the nineteen seventies had been bright with promise as she and her husband considered where in this glorious state they would set up a brand-new janitorial business. Looking at the various cities and towns throughout the state, they picked the Kenai Peninsula, and in particular Soldotna, as a place with a future—a place where her family could be raised in peace and harmony with nature and with the small-town folks around her.

But Leigh's problems at that time weren't with nature, or the people around her.

Her marriage was breaking up. She whisked her children back to Seattle where she could regroup her emotions and her thoughts. She returned and tried to reconcile with her husband, but it didn't work.

She stayed in Alaska and her husband left the state.

During the year that followed, Leigh found a job as a lab technician. That next summer, while trying to take care of registering patients at the counter while her co-worker, Pat, was on vacation, a man in the room caught her eye, but she was too busy to be very concerned. He was good looking, and about the right age. She wondered vaguely if he was single. She had been single quite a while, suppressing that natural instinct for male companionship and instead devoting herself to her job and her children.

She looked up again at Clint and tried to think about where she had seen him before. Had he been in one of her classes? After the divorce, she had signed up to take several college courses in the natural sciences. Had he been on the trip she had taken with the geology class when they traveled to Sterling to observe the old channel of the Moose River to see exactly how the glaciers of the area changed the landscape? She tried to picture her class standing on the hill listening to the professor as he pointed out the landforms. She pictured each member of the class. No, Clint had not been among them.

It must have been at a job. Even though she had been in the medical field for many years, one of her first jobs, after handing over the janitorial business to other family members, was working for the Division of Land and Water Management with the State Department of Natural Resources. It had been her job to plot out people's water rights and learn to read plat maps and other large land maps. Had he been one of the people for whom she had gotten water rights worked out? Had she worked for him in any way? She couldn't remember.

A voice interrupted her thoughts and her work. She looked up to hear Clint talking to her.

"Didn't you work for the state parks a couple of summers ago?"

Leigh's memory flashed on that summer. Of course! He was the man operating the cat when the division of parks was upgrading that old park. She had spent a summer working for state parks through a CETA grant.

Leigh looked up and smiled sweetly. "Oh, yes. I helped brush out the trails after you went through with the cat."

"You seemed to really like working outdoors. How can you be cooped up in an office like this?"

Leigh chuckled. "Well, you work where you get a job. And besides," she shrugged her shoulders, "this is where my training is. I didn't really have any training in making trails."

"Well, you could have fooled me . . . Say, How would you like to get out of this office, and out of town for a couple of days? I'm taking a group of Boy Scouts down the Swanson River and we need another adult chaperone."

Leigh looked at him incredulously, "I've never been in a canoe!"

Clint laughed. "That's O.K. Many of the boys haven't either. It's really going to be a fun trip. We'll go to a lake first to teach them some of the basics of canoeing, then we'll head to the river. The river is really slow. There's no danger at all."

Leigh gazed at this man with slightly graying, light-brown hair, medium build and laughing eyes. This man might be someone she could get to know better. What better way than to go on a trip with a bunch of Boy Scouts. Nothing could happen that way. She would be sure to be safe, at least from Clint! She wasn't sure about being safe from the brown bears or the river itself!

"When are you going?"

"Next weekend."

"Sure, I'll go," Leigh said.

Life took a turn for the better for Leigh, after meeting Clint. He was a kind person who truly took an interest in her as a person, and in her as a mother and woman. He was patient, and believed in her.

They were finally married in 1985. He moved from his trailer into her's in a park called River Terrace in Soldotna. They started talked about their future, and about their retirement. Neither of them was young, but not old enough to retire, for both financial and age reasons.

One day Clint was talking about selling his ten-acre tract of land in the Scout Lake area of Sterling so they could have some money to retire on. Leigh had never seen this piece of property, so they drove out to it. A road went to the property, but that was it. The land itself consisted of a large gravel hill with a drop off down to an aspen grove, then out to a swamp of muskeg.

"It's really a pretty nice piece of property, but potential buyers can't see it, because of this drop off we're standing on." Clint, who operated heavy equipment, shrugged his shoulders in disgust. "I really don't understand it. This whole hill we're standing on is gravel. It'd be easy to take a cat and level this area off and push the gravel down there and build a road to the aspen grove. I kind of thought that would be a nice place to build some cabins and be able to rent them out. Kind of have a little bit of extra income, or maybe give some of these kids who come up for the summer to work in parks a cheap place to live. And see that point over there?"

Leigh nodded, entranced by the scheme of the land.

"We could eventually have a really beautiful house over there, overlooking the swamp and the Kenai Mountains."

Leigh was breathless. She saw fantastic possibilities for the land. She grabbed his hand. "Let's go over to that point, then walk around the aspen grove."

"Well, it's pretty steep."

"Let's go."

Leigh and Clint stepped and slipped down the gravel cliff, then climbed back up the opposite hill. "Oh, let's do it!"

Leigh exclaimed. "You don't want to sell this land! Let's live here! It's beautiful!"

Clint frowned and leaned against a tree. "There's one more problem. Someone once tried to put a well on the property. He went down two hundred eighty six feet and never found water." He pointed to the attempted well site, back toward the other gravel cliff. "It was a dry hole. What do you do with ten acres without any water?"

"Ten acres without water?" Leigh didn't believe it. "There's got to be water here. Back in Oregon we hired a witcher to find water."

Clint looked at Leigh skeptically. "A witcher? I didn't think that was for real."

"Sure they are. Those divining rods really work, at least in the hands of a witcher. Come on, Clint. Let's give it a try. What have we got to lose?"

"The cost of the witcher, for one thing, then the cost of another well if we don't find water. The first one cost $4,500."

"I still say we try it. I'd love to live here. And just think about retiring here. We have to have a lot of water to put in cabins or a camp ground down there in the aspen grove."

Clint finally agreed, and they found a witcher, hired her, and she went to work. She hadn't been walking the land very long when she got a strong pull on the divining rod. She followed the pull and it ran across the land, as if in a narrow stream. Clint sank the well where she first found the pull, and, sure enough, sixty feet down the driller struck water.

But not enough water.

Clint was discouraged, but Leigh didn't want to give up. She hung on Clint's arm and pointed to the aspen grove. "Just visualize four log cabins down there, along with a wash house and a shower house, and spaces for campers. It's going to happen. I know it is. We just have to keep digging. That divining rod had a bigger pull than this measly amount of water would attract."

Clint gave in, and the driller continued his search for water. Suddenly, at eighty-four feet, there was water. More water than they could possibly use. It was cold, so cold, it hurt your teeth. And they could pump out twenty gallons a minute.

"Wow!" Leigh exclaimed. "I think we're in an underground river."

They finished the well at eighty-six feet, built a wellhouse over the pump and wellhead, and locked it up. They were ecstatic. They were finally going to move out of the trailer park and onto their own piece of property in country that was truly Alaska.

Getting married and moving to their own property out of the city gave Leigh just the stability she needed to get involved with other activities in the community. Because of her love of the land, and her desire to see that it stayed pure and clean, she slowly began to read about environmental concerns.

It was during this period that Leigh learned about what goes into the ground as part of oil field waste. And she learned about what goes into the air as part of the refining process. But what she learned about the state's watchdog, the DEC, really disturbed her. She saw how the families at the West Poppy Lane pit had fought to get recognition of their plight for years before any action was taken. And even then the action was half-hearted, with the DEC "requesting" that Unocal clean up the pit, but not truly enforcing the state's law prohibiting the disposal of the waste in the first place, despite the fact that in addition to the benzene plume and the peat soaked with gas-well condensate, drilling muds and buried barrels of propylene glycol were also found at the pit. The DEC did not file any kind of charges against Unocal, flatly stating that they'd rather get the oil industry to cooperate in the clean-up rather than fight them in court. Unocal did not cooperate in the clean-up, so the pit's "silent partner," Marathon Oil, stepped in to take over the clean-up, and is slowly working on it to the publication day of this book, although not to the satisfaction of the landowners, whose property is now virtually worthless. They maintained the company should buy them out and battled that issue in court.

Leigh became disturbed at other areas where she had once assumed the DEC was protecting the public and discovered they weren't. One area she found particularly disturbing was the lack of official knowledge about the peninsula's groundwater. All the DEC officials she talked to insisted groundwater moves extremely slowly. One DEC official stated the water moves only several inches a year, as a rule, so pooh-poohed people's insistence that their groundwater was contaminated by a source several hundred yards away. Leigh's gut feeling was that there were underground rivers.

Thereafter traumatic events effected the oil industry on the Kenai Peninsula. The steelhead Oil Platform, the pride of Marathon Oil, blew out, becoming a torch in Cook Inlet—a virtual tourist attraction, as nothing could be done to hide such a visible symbol of technology still not able to measure up to the forces of nature.

But a more sinister event—one behind the closed gates and out of the public's eye—also occurred, an underground blowout in the Swanson River Oilfield. It was a full week before ARCO admitted to the media there had been a blowout, and then information about the nature of the blowout was extremely limited. Looking back, Leigh believed it was that blowout which was the root of her troubles. ARCO officials openly scoffed at such theories, but never really explained to the public, geologically, what happened in the oil field. They stated that Leigh's own theory was so farfetched it wasn't even worth discussing. The oil field was, after all, located several miles from her house. They theorized that it was impossible for groundwater carrying any kind of contamination to move that far that fast without being discovered in between.

December of that year was a fairly exciting time for Leigh. She and Clint and her daughter, Sue, were now living in their trailer on their ten-acre parcel. Clint had cleared a road into the aspen grove for cabin sites, and had begun to build one of the cabins. The time was marred only by Sue's health. The twelve-year-old had gotten a cold with the onset of winter, and now was unable to shake it. In fact, the cold had developed into a type of bronchitis, along with flu symptoms.

Leigh's oldest daughter, Tessa, who was twenty-three at the time, also came to stay with Leigh, along with her husband and two-year-old son, Neal. While there, Tessa discovered she was pregnant, and it was a joyous announcement for everyone.

One trailer is a small place for that many people to stay for too long, so Tessa and her husband soon moved into Clint's old trailer in the Riverside Terrace Park in Soldotna. But the park lost its well in mid-December, so Tessa started hauling water from Leigh's home. Even after the park's well was restored, Tessa continued to haul drinking water because the water at the mobile park tasted so bad.

Meanwhile Sue's symptoms continued to get worse, forcing Leigh to stay home with her daughter quite often. While at home, Leigh consumed a tremendous amount of herbal tea. One day in late December Leigh was drinking her tea, and noticed it burned the inside of her mouth. She checked out the package and saw it was the same type she usually drank. She dumped out that cup and brewed herself another with a different bag. That tea also burned the inside of her mouth.

When Clint came home that evening, Leigh expressed her puzzlement to him. "The funniest thing happened to me today. While I was drinking my tea, it started to burn the insides of my mouth."

"What kind of tea were you using?" Clint asked.

"The same kind as I always do," she replied.

Clint eyed Leigh suspiciously. "You're not letting all this environmental stuff get to you, are you? Thinking maybe our water is contaminated, too? I don't believe it," he scoffed. "Maybe they sold you a bad batch of tea. Get some more and see what happens."

"That's what I intend to do."

Leigh threw out that batch and bought a new one. But that tea, too, burned the insides of her mouth. She threw out that batch and changed brands. But that changed nothing. Soon she had blisters on the inside of her mouth.

Then some other strange things started to happen. Sue continued to get worse. She complained about her chest, muscles, head and throat. She complained that her lungs hurt her. Then Leigh, while working in a doctor's office, began seeing what she considered to be an unusual number of lung complaints come through the clinic. There were a lot of different kinds of complaints, and in patterns not normal for winter in Alaska. At about the same time Leigh began smelling what she thought was chlorine gas at different places around the Sterling and Soldotna areas.

Events began to pile up on each other. Tessa became extremely sick, and began vomiting in a manner unlike morning sickness. Then the doctor became worried because of the hyperactivity of the fetus. He did a sonorgram, then an amniocentesis. Finally, he broke the horrible news to Tessa. She would have to have an abortion; the fetus was grossly deformed. It had developed what appeared on the screen to be a reptilian-like head.

The shock was devastating. Tessa screamed out to her mother in anguish. She was opposed to abortions, and here was a doctor telling her she had to have one. She had to abort the baby she longed for, growing in her womb.

Mother and daughter cried wretchedly. But the facts had to be faced. The baby was not really a baby. Death was inevitable. Leigh comforted her daughter, telling her she would be able to have another baby right away.

Meanwhile, affliction hit other members of the family. Tessa's little boy, Neal, became extremely fatigued, crying and fussing a lot. Finally, Tessa brought her son over to Leigh's house.

"Momma, I don't know what's wrong. Will you look at Neal? He broke out with these hives, and I was wondering if you've ever seen anything like it?"

Tessa lifted Neal's shirt. Covering his trunk were black, massive raised hives that left him black and blue. They were hideous looking, and Leigh's heart fell.

Tessa continued, "The doctor thought he might have reacted to this medication he was taking, but it was stuff he had taken before and he never reacted. The doctor called them vascular hives."

Leigh was utterly speechless. Some of the events were proving to be too incriminating against Leigh's water supply to continue to totally ignore it.

"You've got to stop drinking our water," Leigh told her daughter.

"Oh, come on, you're not going to start with that again! Our water tastes so terrible, and your water tastes so good!"

"But taste isn't everything. There are blisters in my mouth; you've lost your baby and now there are hives all over Neal. And do you know that both Clint and I feel a burning sensation when we take a shower?"

"Clint does, too?" Tessa asked. The knowledge that someone besides her mother might be feeling some direct effects of the water lent a little bit of credence to Leigh's belief.

"Yes. And he's been having terrible headaches."

"Oh, Momma," Tessa sighed. "Headaches can be caused by many things."

"That's true," Leigh admitted. "But there's getting to be too much evidence against our water. And look at Sue. She still hasn't gotten well."

"I'm still not convinced that's not in her head, Mom. But maybe you're right. We'll start hauling our water from some place else."

In reality, not only had Sue not gotten any better, life had become extremely difficult for her.

"Mom, I can't go to school today. I'm just too tired. All the muscles in my entire body ache. I don't do anything when I'm in school anyway. All I want to do is sleep."

Leigh considered her daughter's statements. The teachers had been concerned that Sue was really having a terrible time concentrating in school. She didn't seem to be paying attention and her grades had fallen off.

"Mom, it's my legs again today. They're cramping so bad."

Leigh shook her head in disbelief. What was she to think? The doctor couldn't find anything wrong with her, and insinuated her daughter was trying to get out of school. Sue had never been like that in the past. She was a good student who looked forward to school. But this was seventh grade, the first year of junior high, and there was puberty and all that to consider. Maybe the doctor was right.

"No, Sue." Leigh said emphatically. "You can't stay home again today. You've got to go to school and do something to pull those grades up."

By now, Sue was almost in tears. "Mom, I try. But I just can't concentrate. All I can do is fall asleep on my desk. I don't know what's wrong."

Anger welled inside Leigh. She believed her daughter. She didn't believe the doctors who said nothing was wrong with Sue. But still, she turned to her daughter.

"Sue, you've got to go. Just do the best you can."

Sue's symptoms continued to get worse. Her stomach problems were terrible, and she had a stringy discharge which came from her bladder. When she would run at school, her legs would give out from under her. She had blisters in her throat, and when Leigh took her to the doctor, he told her it was her adenoids.

"But the kid doesn't have any adenoids!" Leigh exclaimed. "She had her tonsils and adenoids removed a long time ago."

All Leigh could really do is watch helplessly as her daughter continued to experience pain no twelve-year-old should have to.

For Leigh, herself, in addition to all the blisters in her mouth, her uterus started growing, and became as big as a grapefruit. And she would experience headaches and muscle aches and extreme fatigue. Of course, all these symptoms could have been some kind of weird bug traveling through the house. But Leigh didn't believe it. Her own medical background led her to become more and more suspicious of their water.

Finally, Leigh sat down and listed the various ailments she and her family were experiencing: vascular hives and rashes, gallbladder problems, stomach pains, hemature, headaches, substernal chest pain, muscular aches and weakness, hoarseness, congestion, chlorine taste and smell, burning of sinuses, burning of eyes and mucous membranes, acne, nausea and instantaneous vomiting, prickly feeling to the skin, numbness of hands and feet, vision problems, anemia, peeling of mucous membranes, and an erratic heart beat. Later, she added an aborted fetus due to gross deformity, removal of eleven gall stones from her twenty-three-year-old daughter, and her own complete hysterectomy due to tumors.

After reviewing the list, her decision was made. She would no longer allow her family to drink the water from their well. She collected milk jugs and thermoses and hauled them to the fire station in Soldotna and began hauling her drinking water.

But the family didn't stop using the water for other things. They continued to bathe in it, do their laundry, cook with it, and even use it in the steam vaporizer when she babysat her grandson. Sue also got an itchy rash, and the entire family became sensitive to the sun, with their eyes hurting and itching when in the sun.

One day, Leigh noticed an odd odor while doing the laundry. She checked some of the clothes, and, on a tightly knit nylon article, she discovered something that looked like crystals. She changed soap, but the crystals were still there.

When Clint came home, she said to him, "My arms feel like I've been around fiberglass. They feel like they have little slivers in them."

Clint reached out and rubbed her arms.

"They do feel like you've got little slivers on your skin. What kind of soap did you use?"

"The same kind of soap as I always use, but I changed soaps, just to see what would happen, and they still feel the same way. You and everyone else keep telling me it's not the water, so I keep trying to prove that it isn't, but I keep coming back to the fact that it HAS to be the water."

Clint nodded his head in understanding.

Leigh continued, "Come back here and smell the water in the laundry room."

Clint frowned. "It does smell quite weird. Not like diesel, or benzene. But more of a sweet smell. Why don't you get the water tested?"

The next day Leigh called the DEC. They couldn't come out. They were no longer doing individual well testing for free unless there was an obvious problem with the well, such as benzene odor, or evidence of fecal coliform. With the publicity from the West Poppy Lane gravel pit, there were too many people wanting their wells tested just to give them the assurance their water was O.K.

Leigh decided to send a water sample to a lab in Oregon called Ch2M Hill. The company sent her a cooler in which she could send her samples. A lab official explained over the phone that the cooler had custody seals which would ensure that the samples wouldn't have been contaminated before arrival, and if the custody seals had been broken, Leigh wasn't to use the cooler. The custody seals are there so the samples can be used in a court case.

When the cooler arrived, the seals had been broken, so she returned the cooler. The second cooler arrived with the custody seals broken also. She inquired at the post office and at the clinic where she was employed, and everyone claimed nobody had touched it, so Leigh went ahead and took a sample of her well and sent it in.

It wasn't long before a company employee called her.

"I was calling to let you know we did get a hit of toluene in the sample of water you sent in to us. Now a hit doesn't mean that there's very much in there, so I wanted to talk to you about where you live, and how you took the sample."

Leigh breathed a sigh of relief—finally someone, somewhere, would listen to her exasperation over what she believed to be a contaminated well.

"I have no idea where any toluene could have come from," she replied. "I know they had some sort of blowout in a nearby oil field, some kind of failure of an injection well where they put wastes. Then, there's the Sterling Special Waste Site about two or three miles away, and I know there is toluene in there. We've got buried gasoline tanks; we've got all kinds of stuff around. I don't know. I just know we've been really sick, and there's something wrong with our water, and no one will do anything."

The man talked to Leigh a long time, sympathizing with her, and teaching her how to deal with the situation. "To take an accurate sample, you have to run the water for a long time before you take the sample, and you have to sterilize the faucet. Then you have to make sure you don't get any air bubbles, because the dissolved gases can then escape into the bubble, and we won't be able to test for them. After you take the sample, put it in the cooler and send it to us overnight."

Leigh did this, and, while waiting for the results, decided to find out more about what she seemed to be dealing with. Her research took her to a chemical exposure book, *Dangerous Properties of Industrial Materials* (7th Edition, N. Irving Sax and Richard J. Lewis, Copyright 1989), which listed the effects of toluene. Toxic effects are usually a result of inhalation, but it is considered a poison if ingested, affecting reproduction, among other things. When inhaled, the book lists "headache, nausea, eye irritation, loss of appetite, a bad taste, lassitude, impairment of coordination and reaction time, anemia, leukopenia, and enlarged liver,"

A phone call several days later indicated the lab still couldn't run a sample because Leigh had still allowed an air bubble to get into the sample.

She took another sample.

This time the lab was able to run a test. They found something in the water, but weren't really able to identify it, so another set of samples had to be taken directly from the pressure tank.

Shortly after the samples were taken, Leigh visited the lab in Oregon. She was greatly impressed with the cleanliness and up-to-date equipment compared to the state-utilized labs in Alaska.

This time Leigh met the lab employee in person, and he handed her the results.

"We got toluene again," he said. "That's bad. But what worries me even more are these other things. There are a whole bunch of unknowns near the end of the run. It's your heavier substances, chemicals often found in the oil field, that are going to be found at the end of the run. In doing the test, I told the workers we wanted anything identified, so that anytime there was a blip on the machine, it was reported. When we get a blip at a certain number, that number coincides with a known chemical. But we can only identify a certain number of chemicals. There are hundreds and hundreds of chemicals out on the industrial market, and there would be no possible way to be able to identify them. Nobody really knows what they are. They are these long-chain buggers. That's what really worries me. These long-chain buggers. That's what PCB's are. I did get a hit on PCB 12-60, and a hit on benzene."

Leigh was stunned. She was angry, yet she was relieved. She was angry that someone, somehow, had contaminated her well, and she was relieved that she had found the source of her problem. No longer would she have to listen to people telling her all of her worry over her well was in her head. She was also extremely angry that her family had had to go through torture for several months.

While Leigh thought the test results would finally cause people to acknowledge she had a problem with her well, they didn't. The DEC still wouldn't acknowledge that there was anything in Leigh's water, stating that the toluene was probably the result of a sample being contaminated by her hands when she took the sample.

But one of the agency's higher officials, who is known to have an acutely sensitive nose, did stop by Leigh's trailer on one of his visits to the peninsula. He said he definitely smelled something sweet in the water, but couldn't identify it. He added that the nose is one of nature's most sensitive instruments, more sensitive than the instruments used to measure chemicals in the air or water.

Finally, a DEC official came out to take a sample. Leigh felt he was quite haphazard about taking the sample, putting it into his pocket instead of a cooler.

Leigh watched in amazement, and finally remarked, "Hey, look, you got an air bubble in it."

"Oh, that's O.K.," he said. "We get them all the time."

Leigh quietly shook her head in disgusted disbelief, and chalked that test up to zero. But she thought she'd back the test with her own sample. She drove the sample all the way to Anchorage, keeping it on ice the entire time.

Of course, they didn't find anything, and it amounted to Leigh throwing her money away. She was asked if she wanted to send another sample through, but she said to forget it. "I already know I have something in there," she told the lab technician. "You know you got something in your water when it's burning people to take showers, and burns your skin off."

Finding out that there were nasty things in her water did not end Leigh's ordeal. In fact, it was only the beginning.

While she was now convinced in her own mind that their water was contaminated, and had proof to back it up, DEC officials still believed she was just an alarmist because their own tests showed nothing. They still did not believe the test results of CH2M Hill, stating there was probably some kind of leak in her well system, or something wrong with her pipes, or some mistake in the lab or sampling procedure.

The family had stopped drinking the water, but they continued using it for bathing and laundry. It's a lot of work hauling water for drinking, and the amount of work and tedium involved in not being able to use water in the house for anything was more than she wanted to imagine. She reasoned that discontinuing its use completely was probably not warranted because, after all, the tests had not indicated an extremely high amount of contaminant in the water.

After discovering that toluene was in their water, Leigh took the report to Sue's doctor, also her employer.

"I got the test results back on my water, and they found toluene," she told the doctor.

"What's that?" he asked.

"It's an industrial cleaner. You've been doctoring Sue. Do you know what any of the symptoms are of toluene poisoning?"

"No," he answered. "But I'll get the information for you."

He called the poison control center and got the information, and it paralleled Sue's symptoms, and what was happening to all of the family, but of course at a much higher dose than what the water test results would indicate.

Leigh called the United States Environmental Protection Agency. Officials there told her toluene, at a low dose, doesn't cause any health problems. They said she was safe with that chemical in her water, despite the fact it is listed as a dangerous industrial chemical.

Next, she called the State Department of Health. Her water was causing the ill health of her family, and she figured there must be something that agency could do. Officials there told her there wasn't anything they could really do, that it was the responsibility of the DEC. Leigh signed in frustration. She knew what kind of help she was getting from the DEC.

At least, with the removal of her water as drinking water, many of the family's symptoms began to disappear. Sue's congestion began to vanish, then the headaches and nausea started to go away. Leigh's immediate symptoms disappeared, but some of her other symptoms seemed to continue, such as loss of memory or inability to concentrate. Ultimately she did lose her job. Although it was technically because she didn't want to increase her work load, she couldn't help thinking that perhaps it was because she wasn't working up to the quality of her pre-chemical exposure performance.

10

AWARENESS

Leigh Smithen and Dave Brown never really knew each other, but their problems were intimately related as residents inflicted with the pain of pollution, and witnesses to the poisoning of Alaska. They both saw their state—and, more importantly, their backyards—become dumps for the oil industry. Dave Brown sounded the alert. Leigh credits Dave Brown with initiating public awareness.

Despite becoming a public figure, Dave Brown's business was quickly sliding downhill.

Whether the DEC's systematic destruction of Brown's business and the drying up of his cash flow was deliberate or coincidental, it was successful. It also was obvious that the DEC spent a tremendous amount of time and effort studying every detail of Brown's business while major oil company violators continued business virtually unscathed save for an occasional blatant oil spill that couldn't be ignored because it was in view of the public or highlighted in the local newspaper. The DEC posture, Dave said, "could be compared to stepping on a spider while being surrounded by venomous snakes."

While the DEC continued harassing Brown over the operation of the Sterling site, officials also zeroed in on other aspects of his business which had been keeping Brown afloat while the supply of drilling muds to the site was systematically being choked off.

First, the DEC stopped Brown from disposing of drag reducing agent from Alyeska Pipeline Company, after originally allowing it. Second, they stopped him from cleaning up waste oil to sell as boiler fuel. Third, they stopped him from road oiling. And fourth, they stopped him from taking waste at the Sterling site, even though the permit was still in effect.

But these actions didn't take place until after Brown, fed up with four and a half years of nit-picking and harassment, took his story to the newspaper. That's when the real battle between Brown and the DEC heated up to an all-out war.

Three years previously, Alaska Environmental Industries (formerly Mar Enterprises) had invested $127,000 in cash for labor and materials for clearing up waste oil. At that time, Brown had erected a series of tanks in the yard behind his shop, after cleaning the land and putting in a gravel pad and a dike. There were six tanks in all, two seven-thousand-five-hundred-gallon railroad tank cars for primary storage, a six-hundred-barrel rectangular tank, a five-hundred-barrel skid-mounted tank and two one-thousand-barrel tanks. Brown and his employees worked diligently and creatively to put together an apparatus which would clean the oil. As the used oil came in, he heated it for several hours in the first two compartments to around one-hundred-eighty degrees, sometimes adding demulsifiers. This made the bulk of the water and dirt drop out of the oil. When the time was right, he would pump it up into the little five-hundred-gallon tank where he could reheat it to more than two-hundred-twelve degrees, causing it to boil. Then he would open a valve and let the hot oil go down into the centrifuge which thoroughly thrashed it, taking out almost all of the remaining impurities. The finished oil then went out of the centrifuge into the empty compartment below. Tesoro Laboratory tested the finished oil for Brown, since he was mainly using waste crude oil from Tesoro—although used turbine oil from Unocal Chemicals, and used oil from ARCO were also used. The intent was to be able to sell the refurbished oil to Seward Fisheries for use in its boilers.

An April 5, 1982, letter from the DEC to Seward Fisheries indicates the DEC's knowledge and approval of the project [See Appendix Q].

"It appears that Alaska Environmental Industries will soon have the capabilities of refining Tesoro's residual sludge to a state where it can be utilized in your boiler burners. In a recent demonstration conducted in my presence of burning this material in your boiler burners, it certainly appeared that this indeed is possible. It was noted that no black smoke was present in this trial burning. The DEC had always encouraged projects such as this where a waste material can be converted into a useful commodity. In my discussions with Tesoro's environmental engineers and chemists, I am told that in burning this waste there will not be any more of a problem with air contaminants than what is being burned at the present time (crude).

"The D.E.C. has no objection in burning this material in boiler burners. We will expect that the air quality be maintained. We may also want to conduct some air contaminants emission tests in the future, as this is the first time that this material has been utilized as a fuel. If this program is as successful as it appears, I can foresee your firm converting your other facilities to this type of fuel."

Brown signed a contract with Seward Fisheries which involved Brown erecting on their property two one-thousand-barrel storage tanks, piping, heating coils, a filtering system and a lined dike, together with supplying them with a minimum of 250,000 gallons of oil per year containing not more than 2.5 percent water and dirt. The fish company had a law firm draw up the contract. Brown did not have an attorney look at it, and agreed to stand any cost difference between his price per gallon and what Seward Fisheries actually had to pay. That part of the contract ended up costing Brown $42,500.

The fish company claimed Brown delivered a bad batch or two of oil. Tesoro lab reports indicated the oil was okay, so Brown still has no idea how the water content in the oil could have increased between testing and delivering. He believes the fish company left the covers open on top of the tanks, and when tanks aren't full in the winter, steam coils are turned on and there is a great amount of condensation on the inner tank walls. Whatever the cause, Brown believes it could have been discovered and resolved.

But what really destroyed Brown's company was that the DEC, with the United States Environmental Protection Agency, paid Brown three visits on the waste oil cleanup project.

During the first visit, two men from the DEC came directly to the site. The older one, a short, heavy-set man named Dick Haines said, smiling, "Good work, Brown. We're glad to see someone in Alaska is concerned and conscientious enough to start cleaning up waste oil instead of just spraying it haphazardly on the roads or reselling it as boiler fuel with all of its impurities."

A few months later, they made their second visit. The younger of the officials smilingly suggested, "Dave, why don't you apply for a state grant to recoup a part of the money you've invested.

"Nah," said Brown. "Not interested. You know the kind of bureaucratic knots which can be tied for a mere $15,000 grant. Anyway, I'm not interested in letting my competitors know how my apparatus works."

On the third visit, both men appeared again. Brown noticed their faces wore the same grins as they had earlier. For a few minutes they all made small talk, then Haines' perpetual smile faded.

"Brown, I'm afraid we have bad news," he said. "You're going to have to shut down this operation. The sludge Tesoro has been giving you has been reclassified by the EPA as a hazardous waste."

Brown received an estimate from Environmental Services Ltd. that it would cost $95,000 plus one year's time to get the necessary permits to continue to use the Tesoro sludge. A lot of Dave's life savings—$127,000—were in the centrifuge investment, but Dave complied grudgingly and made new plans.

Shortly thereafter, during another surprise visit from Dick Haines, Dave told Haines he planned to open in a few days using oil from Anchorage. There were two road-oiling companies in Anchorage, but they had limited capacity, whereas Brown had five vacuum trucks with a total capacity of 17,850 gallons plus a cabover tractor with a two-thousand-gallon vacuum tank.

"I intend to head for Anchorage with all of my trucks," Dave told the DEC official. "Used crank case oil has been refined once and has never been classified hazardous by EPA. I'll clean it up and sell it for fuel or oil roads with it."

Brown was fairly confident in his plan because of the recent publicity about there being no legal place to take crankcase oil. The DEC had been crying for someone to get into the legitimate business of dealing with it.

Surprise, anger, then plain disgust was Brown's reaction to the DEC's emphatic "No" to his plan. The official to whom Dave spoke told him he might be creating a new form of hazardous waste with his centrifuge sludge.

The drag-reducing agent screw-up began before Dave "went public," when he had no real reason as yet to distrust the DEC.

In 1980, Alyeska Pipeline Co. had contracted with Brown to see if he could dispose of their drag-reducing agent. Brown talked with two field officers in the DEC office. They said he could do a test on the stuff, so Alyeska shipped six barrels of it to Brown. Dave took the barrels to the Sterling site for an experimental burning. He found the material to be extremely flammable. It would only burn half way, then die for lack of air. He discovered that with a good, steady blast of air he could completely burn the contents out of the drum in a short time, with no smoke. Keiser and Moulton each witnessed separate burning demonstrations and seemed satisfied with the results. Moulton issued a permit for the burning, stating that the following conditions be adhered to: "Maximum combustion efficiency shall be maintained throughout the burning period (using auxiliary air flow)" and "Reasonable procedures shall be used to minimize the amount of smoke generated, and no black smoke shall be allowable." The permit for the Sterling Special Waste Site stipulated that Brown "Be allowed to accept other wastes only after prior written permission has been obtained from the department's Southcentral Regional Office in Anchorage or the Central Office in Juneau." [See Appendix R]

After getting written permission from the DEC Soldotna office, Brown had Alyeska ship a couple of truckloads. Dave burned it at the Sterling site legally and without problems. Later, Alyeska shipped Brown approximately two hundred more fifty-five-gallon drums plus a long cylindrically shaped container holding one thousand-forty-two gallons of drag reducer.

At about this time the permit for Sterling Special Waste Site was renewed and the paragraph that formerly appeared in it allowing Brown to dispose of other types of waste with DEC written permission was left out. He received permission from the DEC to burn the drag reducer in a gravel pit under state open burning regulations. He had the cylinder delivered to the property of a Nikiski resident, Neal Tiler, with that landowner's permission.

At the site, Dave watched as Bob Young, one of his employees, forced the drag reducer out of the cylinder with a compressor and into the pit.

"Burn it," he said.

Bob started the burning. At one point the fire started to get out of hand, so Young extinguished the blaze with dirt. Later, he reignited the fire, but a small portion of the material failed to burn because of the dirt in it. They called Neal Tiler.

"Go ahead and cover it up with dirt," Tiler suggested. Young followed his suggestion. That was in April, 1982. Neal Tiler seemed satisfied and so did Brown. In addition to burning the drag reducer, Brown also had a business relationship with Tiler, from whom he bought some equipment and was making monthly payments of $1,032.61. That relationship was going to come back and haunt Dave Brown in a way no one could possibly foresee.

After burning the cylinder full of drag reducer, Brown took the two hundred barrels out to Jim Arness' gravel pit in North Kenai. Jim had a valid DEC septic waste disposal permit which he wasn't using, so Dave made a deal with him. Although Brown wasn't in the cesspool business, from time to time he was requested to dispose of industrial septic waste. He also needed gravel. So Brown paid Arness a dumping fee on every load of septic waste, and Arness gave Brown a good price on gravel. In addition, Arness let Brown store the drag reducer from Alyeska in his pit prior to burning.

Shortly thereafter, two DEC officials, one of whom was Dick Haines, made one of their unannounced visits to Brown's Kenai shop.

It was a gray, overcast, partly rainy day, exactly the same type of day on which previously DEC officials had told Brown he could never burn. Haines said to Brown, "We want an immediate burning demonstration of a barrel of Alyeska drag reducer right there in your driveway."

"But these are unfavorable conditions and I don't have the proper equipment," Brown protested. "If I knew you were coming, I would have been ready for you. To burn that material with no smoke requires a burning fan or fans set at exactly the right height and sometimes it requires the use of a chimney. It definitely needs clear skies, not rainy overcast."

"Well, we're going to require you do it," the other man said.

Dave burned a barrel and, as he had expected, it emitted some smoke.

"Look at that," Haines said, shaking his head. "Now we're going to have to cut off your permit to burn it."

This left Dave with the two hundred barrels in Arness' pit [See Appendix S].

Brown couldn't afford to ship the material outside the

state, so the drums sat at the pit for about three years until they filled with rain and snowmelt, and began to overflow.

Meanwhile, the other drag-reducing material, the residue of which had been buried and forgotten by all except Neal Tiler, came back to haunt Brown. An investigator with the Department of Health and Social Services requested from Brown copies of the payments made to Tiler. Brown was told to either turn over the cancelled checks voluntarily, or under court order. The following letter reveals what happened to Tiler:

> May 14, 1984
>
> Dear Mr. and Mrs. Brown:
> Attached is a copy of the receipt I left with you and your original checks that were written to Mr. ——. As you probably have read in the local newspaper, Mr. —— was prosecuted on 13 counts of fraud. He received 390 days in jail (all suspended except for 95 days); he is required to pay restitution of $4,554.00 and was fined $2,000.00 (which was suspended) provided he has no similar violations for a five year period.
> Thank you folks so much for your consideration and cooperation. It is nice to know that there are good honest people (such as yourself) out there. Have a nice day.
>
> Sincerely,
> Alan G. Humphrey
> Investigator 11

Neal Tiler called DEC and told them what happened on his property. It was an ideal revenge. Because of his call, the DEC rediscovered the two hundred barrels. A few months later, Brown had become the irritant they wanted to do away with. They were sure they had him this time.

The DEC issued Brown a compliance order to get rid of the two hundred barrels of drag reducer. During a meeting with DEC Alyeska, and Tom Adler, Brown told the DEC that he wanted to put on another burning demonstration at the Adler pit, only this time under favorable conditions. The DEC insisted on no more burning demonstrations, but Alyeska, the operator of the Trans-Alaska Pipeline, which had a significant amount of political pull in the state because it was essentially owned by the various oil companies, protested. Alyeska would be ultimately responsible for any waste cleanup. Finally, the DEC consented.

On the appointed day, the weather was right, the equipment was there, and Brown was ready. A small audience gathered. The DEC official present watched closely.

"There's no smoke," he admitted.

Approval was once again given, and Dave burned the rest of it. He also tried to excavate the rest of the material from Tiler's property, but the landowner kicked him off the property. The DEC tried to make Brown ship the material Outside, Brown believes, in another effort to ruin him financially. An interesting twist to the story is that the DEC requested information from Alyeska as to the amount of money Alyeska paid Brown to dispose of the material [See Appendix T].

Brown tried to comply, but he also promptly sent letters to the DEC and to the Kenai Peninsula School District, demanding to know—under the Freedom of Information Act—what the DEC officer's salary was, and what was the salary of his schoolteacher wife.

Road oiling was probably where the DEC really hit Brown and his cash flow the hardest.

Back in January of 1984, a DEC field officer had informed Brown that all oil would have to go to a test lab to be tested prior to any road oiling. Brown answered that that would be no problem, just give him the criteria the oil had to be tested for.

From then on, Betty called the agency at least once every week, hoping to find out what the test requirements were. During this period, the officer kept telling Betty the agency had not yet received the list of parameters.

Brown had not believed it then and he still didn't believe

it. By this time, he totally distrusted the DEC, and believed the delay was a deliberate effort to cause him to lose his contract. Because of terrible dust problems, road oiling on the Kenai Peninsula began on April 25th of every year. On May 14, 1984, the permit was mailed to him, along with the testing parameters, which included every possible contaminant, including polynuclear aromatics, volatile aromatics, chlorinated solvents, metals, total organic halogens, PCB's and phenois. The test cost $1,300 and took the lab three weeks to perform, and the oil failed the test.

The two road oiling companies in Anchorage got their tests done at the same lab for $300 and had their results in three or four days because they did not have to test for every contaminant on the list, in particular, they did not have to test for phenols or polynuclear aromatics.

Of course, the State Highway Department cancelled the road oiling contract. Brown sold all the oil he had on hand to an Anchorage company, where they could use it and make a profit because they weren't on the DEC hit list.

The DEC, between cutting Brown off from cleaning up waste oil and from road oiling, cost Brown's company around $200,000 per year in addition to the losses from Alyeska Pipeline Company.

Then, in 1985, DEC Commissioner Bill Ross visited the Sterling Waste Site with Bill Lamoreaux, DEC's regional director, and Bob Cannone, a Soldotna field officer. Ross and Cannone rode to the site with Kenai Peninsula Borough Assemblywoman Marilyn Dimmick. The following are minutes from the October 29, 1985, meeting of the Borough's Solid Waste Commission:

"Marilyn Dimmick: said that she had gone through the site with the Commissioner and Mr. Cannone and one of the things that Dave mentioned at one point was that the site wasn't big enough to do what they wanted him to do so it is the Borough's obligation to find a better one. She said that she feels that if we could somehow separate the personality (Dave Brown) from all of these, things would clear up a lot faster. She said that her own impression, especially from hearing the conversation between the Commissioner and his employees was that they were more interested in breaking Dave Brown than they were anything else. She said they were really serious, they were asking questions about his cash flow, and how close he was to Chapter 11 and this was about six months ago. She said that she felt that this was totally improper."

"Keith Laurie (an environmentalist officer with Unocal Chemicals): said that from someone who works with DEC from an industry point of view, he is horrified at the way they pick on one person, company, or individual and just target them and nail them and then all of the rest gets ignored because they don't have the manpower to do everyone the same. He said he feels that the DEC has targeted Dave Brown and they are going to do everything to put him out of business."

"Keith Laurie: said they must not have the data at this point because if they did then they would have shut them (Brown) down."

"Marilyn Dimmick: said she thinks the DEC is running a bluff and if the Borough or Dave Brown had fallen for it then the site would be long closed down by now."

11

THE PAST CATCHES THE PRESENT

Back in 1983, Brown had been agitating the DEC by informing DEC officials about the oil companies' illegal dumping, specifically calling about Marathon's spreading of drilling wastes on the road. Dave definitely was not a team player, one who would scratch another's back. He had not, however, "gone public" with his information. But his attitude toward what he believed were incompetent officials possibly "on the take" from industry, or just ignorant of the potential widespread future problems of pollution, was well known. At this point, Dave still wasn't sure just how far his adversaries would go to discredit him and shut him down.

Later that year, Brown and one of his top employees, John Pike, went over to the Sterling Special Waste Site to prepare for well monitoring the next day. Since the wells were only monitored once every three months, Brown routinely pumped them twenty-four hours ahead of sampling so the water being sampled wouldn't be stagnant, but a true representation of the actual well water. As far as Dave knew, the only other man who knew they were going to monitor wells 2, 4, and 6 was Mr. Dough from the DEC.

After Dave pumped up water from Well No. 2, he handed a glassful to John, and took one himself. They both drank the entire glassfuls.

"Well," Brown said, "This tastes like good clean well water to me."

John agreed. They continued, drinking a glassful from each well, which was their standard practice on the day before monitoring.

Each well had heavy, welded steel hood covers over the well casings extending down about a foot. These covers were locked in place with hardened steel master laminated padlocks. Brown had just changed all the locks on all the wells. But he made what he considered to be a serious mistake: he had neglected to file the key numbers from the bottom of each padlock. Anyone in possession of those key numbers could go to any hardware store where master locks are sold and buy new keys.

However, Brown believed no one—except Dough—was aware the wells would be monitored the next day, and no one except the DEC official knew which wells were to be monitored.

The next afternoon Dough met Brown and Pike at the site. As they started up the pump in Well #2, a putrid smell came from it.

Dough shook his head. "Oh, oh," he said. "Hydrocarbons. That's terrible. This will get you shut down."

Dave eyed Dough suspiciously.

"Wait a minute. We pumped these wells yesterday afternoon and John and I both drank water from them and they were fine. Isn't that right, John?"

"Right," John nodded his head in agreement,

Brown continued. "They were O.K. then. If they're not now, it's because someone has spiked them."

Dough shook his head and smiled sarcastically. "Nope.

These wells aren't spiked. That's ludicrous. The site has finally leached into the monitoring wells."

"Oh, is that right?" Dave angrily countered. "Well, I'll bet we'll find hydrocarbons in wells no. 4 and 6, too."

Sure enough, the same putrid smell rose from those wells, too.

"Now let's check the other wells," Dave said. "My bet is that they don't have any smell."

No, you're wrong," Dough said, but he didn't stick around for the other wells to be sniffed. Dave did. He was right. The other wells did not have any smells.

After that incident, Dave immediately bought another set of master hardened padlocks and, this time, filed off the key numbers. Then he changed the locks again and every day for a week John and he went out and pumped down the wells. One week later, the smell was completely gone and they were again drinking the water. Dave could never really prove it in a court of law, but he believed he knew who had spiked his wells and why.

Following the incident, Dave received a letter from the DEC. It stated:

"Because of a noticeable odor at the time of your most recent monitoring at the Sterling Special Waste Site, we were especially concerned about the results. It was primarily for this reason, after failing to receive the laboratory results when due, that I requested in my letter of July 6, 1983 that you cease disposal of wastes at the site.

"We have now received those analyses. In general they are consistent with previous sample results. With respect to oil and grease, however, we believe further analysis and investigation is necessary. Our laboratory in Juneau is continuing to analyze replicate samples. It is possible that when these studies are complete, we will wish to discuss modifications to your operating procedures, monitoring plan, or some other aspects of your permit."

"Modification, huh!" Dave scoffed when he read the letter. "The technical terms of the permit are already impossible to adhere to. They can modify me right out of existence, which is what it appears they're trying to do."

Sure enough, a few months later Brown got hold of a memo sent to the DEC from the attorney general. The memo seemed to Brown to draw a clear picture as to how to structure a modification so it would effectively act as a termination. If the officials were searching for a legal way to shut the site down, they now had it. The memo, dated December 21, 1983, stated:

"You have asked two questions about AS46.030120, which governs the modifications of waste disposal permits. You first asked whether you are correct in assuming that a finding of contamination in surface or underground water would constitute 'a material change in the quality . . . of the waters of the state' under section 120(b), thus authorizing DEC to modify a permit. We believe you are correct. The language is certainly not crystal clear, and there is no relevant legislative history available, but the construction you propose seems consistent with both the letter and the intent of the statute. Of course, the degree of contamination would have to be more than minimal to constitute a material change.

"You also expressed the view that DEC would have greater flexibility if subsection (b) were changed from 'terminate' to 'modify or terminate.' You clearly would have greater flexibility, but we wonder if this change is necessary. Under (a)(1)—misrepresentation or failure to disclose in the permit application—termination rather than modification would seem to be the appropriate remedy. Ditto for (a)(2), violation of the conditions, though if the violation is a minor one DEC might just settle for a promise that it won't happen again. For (a)(3)—material change in the quality or type of water disposed—again termination might be more appropriate, with a new application to follow. At any rate, this sort of material change is likely to produce a material change in water quality, thus authorizing modification under current subsection (b).

"As to (b), DEC might want to terminate, rather than just modify, a permit if it causes water quality problems. However, DEC can condition a permit on the non-occurrence of such problems, so that if they occur, the holder would be violating the conditions, and termination would lie under (a)(2). Moreover, in most cases a modification can be fashioned so as to effectively operate as a termination.

"In short, it is our theoretical opinion (devoid, of course, of any practical knowledge of the realities) that DEC under the current statutory scheme should be able to deal adequately with any problems."

The Sterling site was not closed down as the result of the hydrocarbon odor in the wells. But it put Dave Brown on notice that he was a wanted man by the DEC. Less than a year after that memo, Dave took his information about the polluting oil companies and the eye shutting of the DEC officials to the local newspaper. Less than a year after that, he was told flatly that the state wanted to rid itself of Dave Brown.

Of course, it was in a roundabout manner. No kind of condemnation process like than can be open. However, Brown had a habit of marking important events on calendars, even if the events were phone conversations. On July 30, 1985, Dave was talking to his former attorney, Bill Bixby. Bixby had purchased the Valdez law practice of Steve McAlpine after McAlpine became Lieutenant Governor of the state. Bixby told Brown that, during a conversation with McAlpine on July 25th, McAlpine told him that the state of Alaska was out to get Dave Brown. Brown immediately recorded the date these comments had been made on his kitchen calendar. And the words raised Brown's hackles even higher. And the higher Brown's hackles were raised, the more obstreperous and vocal Dave became, and the faster DEC officials wanted to get him out of their hair. Pretty soon, it seemed a bad joke, when almost a pat routine of events followed obviously designed to rid the peninsula of Dave Brown. Certainly his crusty ways had antagonized plenty of influential people. The oil industry hated him because he had calculated how much and where drilling muds and wastes had been illegally dumped and publicized it. The DEC wanted to get rid of him because Dave kept broadcasting news of its inadequacies and he was a constant thorn in their own side, publicly accusing them consistently of incompetence and collusion with the industry. Many of the residents of Sterling wanted to get rid of him, although they didn't connect that with his public personality. They were just convinced that no matter what the monitoring well results showed, contaminants were leaching from every site in their district. Getting rid of Dave would, at least, be a first step in stopping pollution. They were convinced their wells, and their very lives, were threatened.

The borough didn't like him. Mr. Faultless and Perfect seemed to have a personal vendetta against Brown. The rest of the borough officials felt that Dave was a constant irritant, loudly ranting about the crooked and easily influenced petty politicians in the state.

And lastly, plenty of Brown's competitors didn't like him, specifically Fred Connors. Connors had some personal gripes against Brown, and the vendetta between the two soon turned into a two-ringed circus. Brown had caught Connors dumping waste oil at the Kenai landfill after hours, in spite of the fact that it was illegal to dispose of any oil waste at the site. When Dave caught Connors, their interchange was punctuated by Dave clenching his fists and yelling at the younger, powerfully built Connors, "There is even a sign on the gate stating, ABSOLUTELY NO OIL MAY BE DISPOSED OF HERE."

Brown made the most of the skirmish caused when he had caught his competitor red-handed in an illegal act. He took copious pictures. Then he called the Kenai Peninsula Borough, the DEC and *The Peninsula Clarion.*

It was later alleged that Mr. Faultless and Perfect had given Mr. Connors the key to the landfill. Mr. Dough then trotted out to the site and stated to *The Peninsula Clarion,* "We know what's in there, we have the pictures. Oil and grease is prohibited in the Kenai landfill. It boils down to what is too much. It'll probably be a decision we'll have to make. I hate to see a precedent set where waste from an oil spill comes to a landfill."

Dave thought it less than funny that DEC would state that it boils down to the amount of oil dumped, when any oil is specifically prohibited. He said grimly, "It was the source of the oil which explained the shenanigans."

The oil had come from a clean-up on an old landfill in Nikiski. The landfill was on borough property which formerly belonged to the state. The oil had been there probably over ten years. Mr. Faultless and Perfect said the borough wanted to clean up the solid waste, and some of the solid waste was covered with oil. It was four dump truck loads of solid waste covered with oil that made its way—with the blessings of the borough official—to the dump. Mr. Faultless and Perfect and Mr. Dough's explanations for the illegal dumping were that there was no place else to put it. Without Brown's camera and calls to *The Peninsula Clarion*, the incident would have been covered up, literally, without anyone finding out about it.

Still smarting from this incident, Connors and Mr. Faultless and Perfect wanted a way to take revenge against Brown, and a proposed event dangled itself temptingly in front of them.

Brown's permit at the Sterling Special Waste Site allowed him to dispose of water containing not more than four percent oil. Usually, the oil content was below the four percent. Brown always made a practice of skimming the oil from the surface of the ponds so he could reclaim it for road oiling or fuel. He didn't really worry about the four percent requirement, although he never knowingly took water with more than four percent.

On August 21, 1985, one week after the incident of dumping oil in the Kenai Landfill, Brown got a call from Marathon Oil, requesting that he pump the bilge water from one of the rig tenders. The call struck a discordant note in Brown's mind, because he hadn't received a business call from Marathon Oil in three years. Marathon Oil was co-owner of the Cannery Loop #2 drilling mud pit that Brown had dug up for all to see, and Brown had turned in Marathon's alleged spreading of drilling muds on the road at Beaver Creek Gas Fields to the United States Fish and Wildlife Service and to the DEC. Brown didn't think he was too highly thought of in Marathon circles. Still, his cash flow had been virtually cut off by this time by the DEC and he badly needed the money. He jumped at the opportunity. It proved to be his last leap.

Brown received the phone call at 4 p.m. August 20th. The expediter for Marathon Oil on the Rig Tenders Dock wanted a vacuum truck on the dock at 8 a.m. on August 22nd to pump off a load of bilge water from the Rig Engineer.

In normal pumping procedure, the driver of the vacuum truck stays on the dock to tend his hoses and his valves and gauges, and a crew member goes down below with the hose. Brown had always had to rely on the crew member's integrity. When Dave's driver, Tim Adams, arrived, he pumped the load, then headed for the Sterling Special Waste Site. At the site he saw something alarming.

"Dave," he said in a shocked tone, calling his boss on the two-way radio, "You better get to this site damn quick."

When Brown arrived, waiting for him were Fred Connors and a state trooper. Then two DEC officials pulled in, followed by Mr. Faultless and Perfect, and Anna Toppa from the Borough. They were followed by Polly Crawford, a reporter from *The Peninsula Clarion.* Brown's hackles rose, along with his blood pressure, as he realized he'd somehow been set up. He didn't yet know how, but he did find out that it was Fred Connors who had called Polly Crawford and the DEC.

In normal events, when blowing the load off a vacuum truck, the oil rises to the top and the water is first to drain off. Next comes straight oil, because oil and water don't mix. In the dike at the edge of the pond, there are always two or three erosion holes created by the first blast, and where the load peters out and the stream diminishes. Many of the people at the site were standing around, smirking, while Brown was shouting, "Set-up!"

Ted Warren, one of the DEC officials, had a jar on hand, and stepped gingerly into the mud at the edge of the pond and scooped a jar full of oil that had collected on one of the erosion holes. From that sample, he was going to calculate what percent of oil the entire load was. But he wasn't going to rely only on the jar. Just looking at the sheen, he said to the reporter that he could tell there was much more than four percent oil in the load.

Brown yelled at him, "You can't get a decent sample

from that erosion hole. You have to go to the middle of the pond."

Warren shrugged his shoulders, "But I only have my low shoes on."

The resulting article in *The Peninsula Clarion* showed the comedy of the entire situation:

"A feud is beginning to boil on the central peninsula. Instead of the Hatfields and the McCoys, the players are the local vacuum pumping services, Alaska Environmental Industry and Construction Services Inc.

"Caught in the middle is the Department of Environmental Conservation.

"At stake is the ultimate proper disposal of wastes.

"The latest scene took place Wednesday at the Sterling Special Waste Site, operated by Dave Brown, owner of Environmental Services. As one of his vacuum trucks pulled in to unload its load of bilge water, Fred Connors, owner of Construction Services, a state trooper, DEC official Ted Warren, and Borough Solid Waste Engineer, Skip Bambard, and later a reporter from *The Clarion,* converged to witness it.

"The bilge water that was dumped at the site was much oilier than normal bilge water, and probably far exceeded the four percent oil that is permitted for the site. It came from the Rig Engineer, a boat operated by Marathon Oil to service its platforms.

" 'This is a set up,' Dave Brown said to the group gathered in the rain. 'I haven't pumped the Rig Engineer for years. I blew the whistle on Union Oil and Marathon Oil on Cannery Loop 2 where drilling muds were dumped without a permit. They wouldn't do business with me. I should have smelled a rat.'

"Connors was behind the calls to the troopers, the DEC, the borough and *The Clarion.* He said Marathon Oil called him Tuesday afternoon to pump the Rig Engineer, and he turned down the job. He said he used to haul bilge water but doesn't anymore. But he added that he knew Brown would accept the job, so went to the boat Wednesday morning and then followed Brown's truck.

"While he wasn't sure the load was oily, he said, 'I felt like there was oil in the truck, so I called the troopers—as witnesses.' He added that if the truck would have pulled into Brown's shop along the Spur Highway, he would have discontinued his pursuit, because he would have assumed they were going to store the water until the oil separated out, which he claims is the proper way to dispose of bilge water.

"But since the truck continued, he said he knew it was going to the Special Waste Site, so made the phone calls.

"When the truck was emptied, Connors admitted that the water was much oilier than normal bilge water, but said that even normal bilge water contains more oil than the four percent allowed to be disposed at the site.

"Brown disagreed. He maintains normal bilge water just contains a trace of oil, well within the allowance.

"The incident comes on the heels of an incident over the weekend involving Construction Services, in which Brown contacted local reporters concerning the dumping of oil in the Kenai Landfill. DEC was ultimately contacted, and is investigating that incident.

"This incident is also being investigated by DEC, so Warren would not comment on it.

"But he did say the feud can be beneficial for DEC and for the proper disposal of wastes. 'This is good for us. They're watching each other. We need more eyes. We don't want setups, though. That's counter productive.'

"One question that remained unanswered Wednesday was Marathon Oil's involvement, and if neither Brown nor Connors has been pumping the bilge water from the Rig Engineer, who has been pumping it, and where, if there is no legal place for it to be disposed of, is it being disposed?

"Jerry McLay, the Marathon Oil expediter who called Brown, refused to answer any questions."

While the incident could have been shrugged off as a setup, and feud, it wasn't, and instead of being a comedy, it put the final nails into Dave's coffin as a businessman in the state of Alaska—at least in the area of waste disposal.

The DEC called a hearing to revoke Brown's permit.

The morning of the hearing was cold and damp. Dave and Betty rose early drinking their coffee together. As he dressed, he looked around uneasily. He felt ill at ease in his only suit and began twisting his tie. Courtrooms didn't suit him; he wished he were going hunting or fishing.

Betty reached out to pat his arm.

"Dave, try to stop fidgeting," she said, watching her husband intently.

"I am trying," he said, and made a self-conscious, sweeping gesture. "This is the result."

They both had to laugh in that shared, secret way long-married couples have.

They had agreed Dave would go to the hearing with his attorney. Betty could not afford to miss work. They needed every cent coming in. Worriedly, she kissed him goodbye. "Call me," she whispered. He tried to smile reassuringly. A few minutes later sitting down next to Bill Bixby in the courtroom, Dave's smile faded.

One of the DEC officials, Ted Warren, told the hearing officer that the bilge water at the Sterling site contained more than four percent oil, and that "I took my samples from the pond itself." Warren also testified "After the vacuum truck blew off its load there was an oil slick or sheen two to three inches thick on the surface of the pond."

Dave's attempts to reveal the truth were silenced.

At the end, the hearing officer revoked Dave's permit to operate.

In the room, there was momentary silence. The verdict had hit him like a mortal shot.

Nothing in Dave's experience of life had prepared him for this stunning moment of defeat.

After the hearing, Dave was so irate over what he considered to be the outright lies told about him that he marched straight to the Anchorage Police Department to charge the DEC official with perjury. The Anchorage police told him that since the incident occurred at Sterling and involved state regulations and a state permit, the perjury charge should be filed with the Alaska State Troopers. The state troopers said that since the alleged perjury was committed within the boundaries of the city of Anchorage, it was a city police problem. The complaint got lost in the shuffle.

Brown's angry complaint states:

"The load in question contained sixty-five barrels. In the oil industry, one barrel equals forty-two gallons, so we had approximately 2,730 gallons. Four percent of 2,730 gallons means that 109.2 gallons of oil would be allowable under the terms of my permit. One hundred nine gallons of oil would have quickly covered the entire surface of the pond. Actually, there was a very minor oily sheen between the discharge point and the southeast corner of the pond, perhaps thirty gallons in all, including the two gallons in the erosion hole.

"As to item C, Mr. Warren's lie about the two or three inch thick oil slick or sheen, I have enclosed a statement from a registered engineer together with a second sheet containing his calculations. Any Anchorage engineer can confirm the accuracy of Mr. Ellis' calculations."

Walter Ellis' statement says: "I, Walter E. Ellis, am a registered Mechanical Professional Engineer, registry number 14805, State of Ohio.

"I have personally inspected the waste water pit in the north west corner of the Sterling Special Waste Site at Sterling, Alaska, on July 28, 1986.

"The pond bottom appeared level and rectangular measuring 95 feet 6 inches by 61 feet 6 inches. This level pit bottom area was confirmed by the existing oil stains.

"My calculations, attached, indicate that a discharge of 60 barrels or 2,520 gallons would raise the level of the pond 0.6436 inches. To get an oily slick or sheen 2″ deep would require 7,831 gallon. To get an oily slick or sheen 3″ deep would require 11,746 gallon."

In layman's terms, the total volume of what was in the

vacuum truck could only have raised the pond level by a fraction of an inch, so that it would have been totally impossible to get an oil slick of several inches as a result of that load [See Appendix U for further calculations based on the area and volume of the pit.].

After all this happened, Brown decided he'd better get an idea about what was in his file at the DEC. After all, it was public information. He spent two or three days in the DEC office paying for and making copies of documents. Finally, Ted Warren told him that his secretary was too busy on DEC business to help him and that DEC needed to use its own copying machine. Brown asked if he could bring his own copying machine to the office and make his own copies. Warren gave him permission and the secretary showed Brown which filing cabinets' drawers contained the information he wanted.

The next day, in the front lobby of the DEC office, Brown started making copies, right under the eyes of two secretaries with two DEC officials present, and with their permission.

In the midst of all the Sterling Special Waste Site folders, he stumbled onto a large manila envelope which said on the corner "Confidential Attorney/Client Communication."

Brown knew what that meant. But, spurred on by his belief he had been railroaded by the state of Alaska, and egged on by the knowledge that if DEC officials were stupid enough to let him into their files and leave in the files confidential information, then he should take advantage of the opportunity. Brown grabbed it.

The manila envelope contained all the information the DEC apparently had on the oily bilge water incident. Brown copied every document, then put the envelope back in the filing cabinet. The following document is probably the single most incriminating piece of evidence in the entire file—the statement of Anthony Zenk, dated August 21, 1985, wherein Ann Toppa, secretary to the Solid Waste Engineer at the Kenai Peninsula Borough, called to tell a DEC official that the "deal happened."

The statement: "I, Anthony Zenk, received one call

from Ann Toppa that said there will be a truck at the Sterling Special Waste Site at 10:00 a.m. and to tell Mike that the "Deal happened." I informed this person that Mike would return at 1 p.m. This call was received at about 9:30 a.m.

"I received a call from Fred Connors at about 9:45 a.m., as I was answering the phone, and I don't recall who he asked for at first, but he did say that an oil truck would be at the Sterling Special Waste Site at 10:00 a.m. and that someone from the DEC should be there to take a sample. He then asked to speak to Ted and said that it was very important. I put him on hold, described to Ted what was said, and returned to the phone and told Fred Connors or person calling, that if he would hold Ted would speak with him. He held for awhile, about five minutes, and then he called back on a different line, then described that he had been cut off, but the light was still blinking. He mentioned that he was still interested in speaking with Ted, I put him on hold again. In about two minutes as he was holding, I then intercepted and told the caller 'Fred Corronors,' that Ted was busy and would call him back if he did not want to hold. Caller, 'Fred Connors,' then said someone should investigate and hung up." [Appendix V].

Also included with the papers was a statement by Connors to DEC officials, part of which states that he knew the bilge water couldn't be dumped at the Sterling Site, and that he had suggested to Marathon that they call Dave Brown. His reasons were, "I decided that due to things that's happened in the past between Alaska Environmental Industries, Dave Brown, that he has pulled on us through articles in the paper and such as, it was time to see if we couldn't do something back." [The first page of his testimony is included in Appendix W]

Nevertheless, his permit revoked, his cash flow cut off, Dave Brown had to declare bankruptcy.

Dave and Betty lost their business, their shop building, their dump site, all of their trucks and equipment. They lost their home, Betty's car, and Dave's pick-up truck. They lost their credit rating, their reputations and their self-respect. They lost all of the fruits of a lifetime of hard work.

A short while later, Dave and his wife were forced to leave Alaska.

PART THREE

12

BETWEEN THEN AND NOW

Taking their few possessions with them, Dave and Betty Brown returned to Seattle poorer than when they had first left over twenty years before.

When Dave had been forced out of business, the Browns hadn't been rich, but they'd had a comfortable life and everything Dave had ever wanted.

Now, at sixty years old, Dave had to begin again, find some sort of employment and pull his and Betty's lives back together. He had no job, no money and no prospects for either. For a while, he couldn't shake his depression, anger and frustration, but, with Betty's support, his desperate feelings finally passed and his fighting spirit resurfaced.

Within a month, he'd found a job; in six, they'd begun saving their money. Soon thereafter, he began plotting his return to Alaska. As he waited and saved, he continued his research and began to record all the reports of the malignant pollution that had gone on and was going on in the state he loved.

Gradually, the malevolent shape of the entire pattern cleared and solidified in his own mind.

He traced his own awareness of that form back to November 1, 1981, when Marathon Oil had spudded its well known as Beaver Creek Unit #6. The well had been completed on September 5, 1982. Dave Brown had hauled much of the drilling mud to the Sterling Special Waste Site from that unit, and later Marathon had cleaned up an old mud pit, taking much of the mud to the Sterling site in dump trunks.

Dave had also done some work for Marathon on its offshore drilling platform, the Dolly Varden, and had assisted Marathon in the cleanup of an oil spill on the road to Beaver Creek in the Kenai National Wildlife Refuge in November 1981.

When Dave was at Beaver Creek drilling site, he had asked Tony Swan the dimensions of Marathon's reserve pit where the mud was stored. Tony, the drilling superintendent, had let Dave know that was none of his business. His quick refusal had sent up warning flags in Dave's mind. If Marathon had intended to obey the law, then there would have been no reason to hide the dimensions of the pit.

Dave had later found out that Marathon had spread most of the extra drilling mud from that reserve pit on the road, including the bridge that crosses Beaver Creek, causing some mud to fall into the stream. Beaver Creek is a small stream that drains into the Kenai River, and is fished mainly by children. The stream contains a small run each year of king salmon, plus some pink salmon, rainbow trout and Dolly Varden.

The mud manipulation had been a four-day long, twenty-four-hour-per-day covert operation, using two or three vacuum trucks and two graders belonging to two of Dave's competitors from Kenai.

During the operation, all workers in the field had been required to enter and leave by car pool to minimize the number of vehicles traveling through the muddy mess. The vehicles had all been washed before passing out of Marathon's locked gate.

Dave had heard the operation was a twelve-hour per day one, but was informed differently when one of the competitors had stood in Dave's office and bragged to Dave and Betty that the operation ran twenty-four hours, not the twelve hours Dave had thought. He also had told Betty and Dave that Marathon Oil was guilty of as much illegal drilling mud disposal as Union Oil.

Once again, warning flags had flashed in Dave's brain. He knew that couldn't be true. Marathon Oil hadn't been even one-fifth as active in drilling as Union Oil. After the competitor left, Dave had turned to Betty.

"I smell a rat."

Betty had replied thoughtfully, "Maybe he's trying to set you up. Maybe he wants you to lose any business you do have with Marathon."

Dave hadn't bit the bait, but he still believes Marathon was remiss in keeping the environment clean. The Beaver Creek pit #6 had an overflow ditch designed for one thing: in the event of a large amount of rainfall, the muds would trickle unnoticed out into the Kenai National Wildlife Refuge lands—lands belonging to the people of the United States—instead of onto the drilling pad where they would cause a huge mess.

While Dave hadn't taken out a newspaper ad about Marathon dumping mud illegally, he had informed the DEC of the spreading of the mud on the roads. He had used his own name, and the result was that Marathon and its drilling contractor, Brinkerhof Drilling, had quit doing business with him.

Then Dave had decided to try a little trick. He reported the incident to the United States Fish and Wildlife Service using an assumed name, "Ernie Cunningham." He had told no one, not even Betty, that he had reported the incident. So when Marathon and Brinkerhof had accused Dave of turning them in, it hadn't taken a wizard to figure out where they got their information. No one, except the DEC had known that he had furnished the information. It was no wonder that people were afraid to call the DEC with complaints about polluting oil companies.

While the DEC had told Marathon and Brinkerhof who had turned them in, no one—neither the DEC nor the United States Fish and Wildlife Service—had done anything about the illegal spreading of mud. DEC had sent two officials to investigate after the operation was over, and the investigators had been told the matter would be handled. It hadn't been.

Marathon did apply for a permit at Trading Bay, a waterfowl-rich delta area across Cook Inlet accessible only by air and oil tanker. The permit, which had been issued by the DEC was to dispose of all wastes from crude oil as it was cleaned and processed for shipping. Disposal was to be in an unlined pit, but there was also permission to landfarm the wastes. Oil and grease had begun to appear in the monitoring wells. The permit expired, and, although the lay person would have had trouble checking it out, the operation had appeared to close.

Marathon operates two offshore drilling platforms, the Dolly Varden and the Steelhead, in the Cook Inlet, in addition to the onshore field at Beaver Creek. The Steelhead has been the pride of Marathon, being the largest and most modern rig in Cook Inlet. The rig, however, has been beset by problems from the very beginning. Its first catastrophe occurred when a crane crashed on the rig's deck. No one was injured, but the result was a twisted mass of steel. It lay idle for quite a while until finally the company began rebuilding the mess. Then, late in 1986, the platform had a gas blowout and served as a torch for Cook Inlet for many weeks. That was finally rebuilt, and the rig is currently operating. However, like many other rigs in Cook Inlet, it has its spills, which go unnoticed and unreported to the DEC and workers, fearful of losing their jobs and being blacklisted in the industry, keep their eyes and mouths shut.

Marathon has had several other public problems on the Peninsula. The company had been the "silent partner" with Union Oil in ownership of the Kenai Gas Fields, The Cannery Loop Fields and the adjacent West Poppy Lane gravel pit.

Since DEC had been having a hard time getting Unocal to cooperate in cleaning up its own mess, and the publicity had been hurting both Unocal and the DEC's wishy-washy methods for relying on industry to clean itself. Marathon officials had stepped in, public relations employees in hand, to "take over and get the mess cleaned up." They were perceived by some as the shining knights out to do a good deed. They were perceived by others, particularly the landowners in the West Poppy Lane area, as surface do-gooders, out to improve their image with as little expenditure as possible. Glossy pamphlets about how the clean-up was coming had kept news media at bay, and the landowners' cries began to get weaker, drowned out by the orchestration provided by Marathon and the DEC.

But all was not totally well with Marathon. Marathon's land manager, Tom Brooks, would meet privately in DEC's Bill Lamoreaux's office, and no one really knew what was going on.

A tiny legal ad told a partial story. As required by law, the DEC had advertised for public comment on a proposed lined, state-of-the-art waste disposal pit to be built on land adjacent to the Tesoro Refinery in Nikiski. A notice, which was subsequently lost, had also been sent to the Kenai Peninsula Borough. The landowners adjacent to the property or in the nearby neighborhood were never personally notified.

This was in 1988. Dave Brown was no longer in Alaska, but he had raised the peoples' consciousness level to near hysteria when it came to oil disposal sites. People were clamoring for clean-up, and for action against the slaughter of the land. The mud-pit pocked landscape was in everyone's eyes daily. A tanker had trashed the inlet by dumping part of its load of crude into the inlet. PCBs were in refuge land. The West Poppy Land pit had been in the news for three years. Governor Steve Cowper had visited the peninsula, personally touring the areas where clean-up was necessary, and the media had recorded the stench and the sickening display of industry gone bad.

To think that DEC officials in Anchorage could slip a

major oil industry disposal site by the people of the peninsula was incomprehensible, but it had almost happened.

The day before the comment period was to expire, one of the residents of the area had noticed the ad. A call had been made to the local newspaper, and the word had spread wildly. A quick meeting had been called. Angry residents had swarmed to the meeting, and the location had had to be changed to a more spacious new high school.

The outcry had been tumultuous. People had been angry that Marathon was trying to put a waste pit, which would have taken waste from the West Poppy Lane gravel pit along with drilling muds and other assorted oil industry wastes, in their own backyards. But they had been more angry that the attempt had been made to do it without holding a public meeting.

Marathon official Tom Brooks, in spite of the catcalls and harassment, had doggedly supported his pit. He had believed in the state-of-the-art system of liners and collection systems designed to absolutely prevent any type of leaching from the site. He also had known that the groundwater in the vicinity of the pit was already horrendously polluted with diesel from the Tesoro Refinery, and the neighbors behind the refinery were already in trouble if they feared for their well water.

DEC official Bill Lamoreaux had been caught in a trap of his own making. All he could claim is that although he had had the authority to call a public meeting, he hadn't, because he hadn't thought having a waste pit in a neighbor's backyard would have caused so much concern. The pit was eventually denied—not by the DEC—but by the Kenai Peninsula Borough. As of the date of the publication of this book, Marathon still had no place to put the contamination exhumed from the West Poppy Lane gravel pit.

While Marathon operated within the Kenai National Wildlife Refuge, the operators of the main field, the Swanson River Oilfield, were Chevron and ARCO.

Oil had first been discovered at Swanson River in 1957 by the Richfield Oil Company, which later became Atlantic Richfield, which eventually became ARCO. The discovery well had been known as Swanson River Unit 34-10. This well had been completed on August 24, 1957.

On November 10, 1957, the Swanson River well known as Swanson River 331-27 had been spudded in, and it had been completed on February 17, 1958. At some point during the drilling of this second well, or immediately thereafter, the operation of the Swanson River field had been taken over by Standard Oil Co. of California, which eventually became Chevron.

A total of one hundred nineteen wells had been drilled in the Swanson River field in what was known as the Soldotna Creek units and the Swanson River Units.

Of these one hundred nineteen wells, thirty-one had been drilled since state solid waste disposal regulations went into effect in July, 1973, yet in all twenty-eight instances the waste drilling mud had been disposed of in unlined pits at Swanson River with no disposal permits granted or even applied for, and in total violation of the law. Disposal pits from all one hundred nineteen wells are unlined and uncapped and are leaching into the ground and into the water table. Unfortunately, ninety-one of those wells had disposed of their spent drilling mud in unlined pits before there was any law that said they could no longer do this. But the laws didn't matter anyway.

Chevron had applied for and received six oil and gas wastewater injection permits, but, although dirty wastewater could be deep-well injected, heavy drilling muds could not be. As earlier suggested, deep-well injection may be the cause of contamination of the artesian aquifer, just as surface pollution from mud pits can cause contaminations of the unconfined aquifer.

Surface mud pits can cause other types of damage, too. In one of his newspaper ads, Dave Brown said, "If you think our local moose all die of old age, highway accidents or hunting accidents, then you are wrong. A lot of them die by falling into drilling mud pits."

Dave had been called on that comment by a Chevron attorney at a Chevron permit hearing in Soldotna. He had antagonistically asked Dave if he had personally witnessed any moose falling into mud pits.

Of course, Dave hadn't personally witnessed this, since the pits were in the fields which were locked to the public. But Dave had known several different oil field workers who had worked in the Swanson River oil fields for many years and who had relayed the stories to him. One specific story had occurred in 1983. The employee had said that when he had gone to work in the morning, a big cow moose had been up to her neck in the drilling mud pit. To get her out, a winch truck had been backed up to the pit. A loop on the end of the winch cable had been thrown over the neck of the cow moose. The cow had then been cranked out of the pit on the winch cable, of course choking the cow moose to death. In order to get their cable back, the workers had had to cut off the head of the moose. Dave had heard this same story from two different workers who had been there.

While it had been apparent to Dave Brown that the DEC officials had different standards for him, a private individual, than they had for the oil industry, those double standards had even applied to the United States Fish and Wildlife Service.

In 1982, Dave had been attempting to find a way to clean up or dry up the two long, deep drilling mud pits on the south end of the Sterling site. The first thing he had needed to do was get rid of the water in both pits so he could get down to the drilling mud itself. Dave had owned five vacuum trucks at the time. The smallest truck had held eighty barrels, or 3,360 gallons. The largest truck had held ninety barrels, or 3,780 gallons. The other trucks had held eighty-five barrels each, or 3,570 gallons each.

With that kind of capacity, Dave had figured he could get rid of the water quickly if he could find a place to put it. While the water on the top of a drilling mud pit might have had an oily sheen on it, the heavy metals which caused contamination would probably be on the bottom. Dave was in the road oiling business, anyway, so he had called Bob Ritchie with the United States Fish and Wildlife Service in Soldotna to see if he could spread the water in the gravel roads in the Refuge, to act as a dust suppressant.

Ritchie had been emphatic. "No!" he had replied. "We are not allowed to put anything on the roads, not even clean water."

But apparently that standard hadn't applied to the oil industry. In 1983 and 1984 Chevron had laid 70,000 tons of oiled gravel on top of many of the twenty-five miles of road around the Swanson River oil field as a dust suppressant. The gravel, it had been reported, had come from a stockpile which had been contaminated with polychlorinated biphenyls (PCB's) after the explosion of a compressor in 1972. By 1985, PCB's had been showing up in samples of the ground taken by Fish and Wildlife officers, and the clean-up had begun. It is continuing to the publication date of this book, and the cost could run as high as forty million. ARCO took over the operation of the field, so is now in charge of the clean-up.

While operating the field, Chevron had applied for two drilling mud disposal permits. One was at the Swanson River unit, and the other was located across the inlet at Beluga.

By Chevron's own testimony, the issuance of the permits may not be in accordance with Alaskan statutes. Chevron had stated that at the Swanson River pit, the water table was "four to six feet" below the lowest point of the proposed pit. Under the new Alaska solid waste regulations, which had gone into effect in 1984, the statutes stated: "A minimum vertical separation of six feet is kept between the waste application surface and the ground water table." It did not say four to six feet.

Chevron's disposal pits across Cook Inlet are located on land owned by the Kenai Peninsula Borough, and Brown believes the issuance of that permit was a violation of the economic viability clause of his lease, which had guaranteed that all drilling muds to be disposed of on Borough-owned land, were to be disposed of at the site. The permit had been issued before the site's closure, but during the period of time when the DEC had been contemplating closure.

Kenai Peninsula Borough Mayor Don Gilman had disagreed with Dave on the interpretation of the economic viability clause long before the question arose. As far back as 1978, the mayor had written, "The Borough interprets the economic viability clause to mean that the borough will not authorize any other private party to operate a similar waste disposal operation on borough lands in the central peninsula area which would deprive Mar Enterprises of a return on its investment as contemplated in its bid." However, the words "central peninsula area" were not in the lease itself.

Stan Thompson had become the borough mayor after Gilman, and Brown had approached him with the problem. Stan had said the issue was Brown's problem, and he wasn't going to get involved. Thompson had suggested Brown might have to sue the borough if he really wanted to push it, so that is exactly what Dave did. That lawsuit is still pending.

Brown has kinder words for ARCO than for the other oil companies which operated on the peninsula. He had been on hand through the drilling of three wells by ARCO, all within the Kenai National Wildlife Refuge. They were Wolf Lake #1, completed on November 12, 1983, Funny River, completed on March 20, 1984, and Wolf Lake #2, completed on February 23, 1985.

Dave had always been impressed that ARCO had seemed more inclined to obey existing disposal statutes and regulations than did any of the other oil companies. The company hadn't appeared to be trying to get away with things, covertly covering its operation. Instead, the mud had been pumped from the pits and disposed of at the Sterling site, or else a permit had been applied for and issued.

While the oil company had appeared to be trying to obey the law, there was still no real watchdog making sure it was. When the entire operation at the Funny River well had nearly been completed, and Dave had only a few more loads of drilling mud left to haul from the site, the DEC's John Dough had asked Dave for directions to get to the well. Dave had told him to follow one of his vacuum trucks to the site, but couldn't help wondering what kind of protection the DEC was giving the environment, when, three months after an initial inspection should have been completed, the field officer hadn't even known where the well was.

Dave had also discovered that there might be very different DEC standards for different oil companies. When ARCO had drilled Wolf Lake #2, the company had applied for two permits—one to inject the light liquids down the annulus of its own well casing, and one to leave all the heavy muds in a lined reserve pit.

This had been done before the DEC had ever enforced any of its own statutes, but was after Dave had taken his issue to the newspapers. ARCO had applied for an injection permit and, while the application was still pending, it had injected drilling liquids into its own well. The company had been cited by the DEC, and had been fined. ARCO had admitted its error and paid the fine.

At approximately the same time, Union Oil had been caught injecting drilling fluids down its injection well in the Kenai Gas Field, after its injection permit had expired and before it was renewed. Union Oil was not fined.

ARCO has a history of more openness, unlike the other oil companies. When ARCO had taken over clean-up of the PCBs, several public hearings and open informational meetings had been held, and the press had been given a tour of the site and information as to how the incineration of the PCBs was to take place. When the company had spudded in a well known as Moose River #1, adjacent to the Sterling Special Waste Site on June 23, 1988, the press had been invited to view the rig, and the lined reserve pit.

Hearing about ARCO's well adjacent to the Sterling site had brought many conjectures to Dave's mind. He couldn't help but wonder if there had been a connection between the DEC's closure of the site and termination of the lease, and the fact that ARCO had wanted to put a well in at the site, looking for the source of the Swanson River oil. The company, however, hadn't found oil and the well had been abandoned on September 17, 1988 after drilling 12,328 feet.

13

HERE AND THERE

The more Dave read and cogitated, the more he learned and remembered.

Union Chemical Company in Nikiski was owned by Unocal—Union Oil of California. Today, the company is known as Unocal Chemicals Division. The plant had been built in the late 1960's, and, at that time, it had been known as the Collier Carbon and Chemical Company. Unocal Chemicals Division is the world's largest manufacturer of urea and ammonia fertilizer and the largest employer on the Kenai Peninsula, employing about three hundred people.

The plant had not always been friendly to the surrounding environment. It had been rated by a national environmental group as the fortieth dirtiest industrial plant in the country. Across the highway from the plant were several acres of dead spruce trees. The plant manager had said that no one had ever really studied what had caused the death of the trees, but he had speculated they might have been "over fertilized." The trees had been cut down in the mid-1980's.

The groundwater beneath the plant had been contaminated for many years with sulfinol, ammonia and arsenic, and a black ooze leaches out of the bluff adjacent to the plant, and creeps down the gravelly beach into Cook Inlet.

In 1974, Union Chemical Co. had made an out-of-court settlement with the State of Alaska which had accused Union of violating the terms of its National Pollutant Discharge Elimination System (NPDES) permit. The company had paid the state $429,000. The state had alleged Union had discharged into Cook Inlet greater quantities of pollutants than the permit allowed, and then had falsified the pollution monitoring reports which were submitted to the state.

In response to a Freedom of Information request, Dave Wilkerson, resource planner with the DEC, had stated in an October 10, 1988 letter, "With reference to the Union Chemical settlement regarding waste discharge in excess of permit limits, which occurred in the early 1970s, it was an NPDES permit, # AK-000050-7. The total volume discharge was approximately 1,000,000 gal day, and there were specific limits for ammonia, organic nitrogen, and oil and grease. The maximum daily limit for ammonia was seven thousand pounds day, and Union discharged a maximum daily discharge of 14,890 lbs. of ammonia. At the time the violation was discovered, apparently Union Oil had been discharging ammonia for a period of a couple years at approximately twice the allowed discharged volume . . . the settlement occurred in 1974 and the violations took place during 1972 and 1973." Wilkerson had added that in addition to the one million gallons permitted to be dumped daily into Cook Inlet, the company was permitted to dump 22,000 gallons of pollutants per day on the ground at the plant.

Wilkerson had enclosed a newspaper article with the letter. The article had been written by John Greely with the Associated Press:

"The state of Alaska is investigating possible criminal and civil charges against Collier Carbon and Chemical Corp., a major producer of fertilizer sought by farmers in the western states.

"A warrant allowing state agents to confiscate a vast amount of company records shows the probe centers on possible violations of a state permit for discharge of ammonia into Cook Inlet from the fertilizer plant at Kenai.

"The investigation comes as the Senate Commerce Committee is considering legislation to allow the export of fertilizer from Collier's Alaska operations to the Pacific Northwest. Collier is a subsidiary of Union Oil Co. of California.

"The warrant alleges that documents seized in a raid Wednesday indicate that Union and Collier officials prepared the false pollution monitoring reports submitted to the state.

"Three possible felony charges, each carrying maximum penalties of two years in jail and fines ranging from $2,000 to $10,000 are listed in the warrant. Another charge under investigation—adding to pollution of Cook Inlet—is a misdemeanor providing a year in jail and a $25,000 fine for each violation.

"The warrant connects four Union and Collier officials to the investigation."

No jail time has been served, and the matter has been settled out of court.

Dave Brown had done a lot of business with Unocal Chemicals. He had averaged $300,000 in business per year, most of it in vacuum trucking at $80 per hour. Some of it had been in rental of equipment, power washers, and tanks. A substantial portion of the money had come from dump fees at the Sterling Special Waste Site.

When Ray O'Docharty and Dave had first taken over the operation of Mar Enterprises and the Sterling site, they couldn't help but wonder how Tom Lyan had ever got solidly entrenched at Unocal Chemicals when there had been other companies on the Kenai Peninsula that had vacuum trucks, and would have pumped for $70 per hour instead of $80 per hour.

They had begun hearing rumors of payoffs to a local bar, from where the money was to be delivered to someone at Union Chemicals. Dave could never confirm the story, but he had found an interesting cancelled check from Mar Enterprises made out to a local bar in Kenai for $4,955. The check had been labeled "commission." It had been dated January 9, 1978 [See Appendix AAA].

During the years Dave and Ray had done business with Unocal Chemicals, Dave had been concerned over the amount of contaminants they had been allowed to dump into Cook Inlet and on the ground, but had decided that would be a problem for the permitting system. But aside from the amount permitted, no one, except a few employees at Unocal, truly knows how much pollution is actually going into Cook Inlet. The level of pollution is monitored by the company itself, not by any watchdog agency.

Dave had also been concerned about the number of sulfinol spills, some of them major, the plant had been having. The spills may have been the result of human error or equipment failure, but they had been frequent—too frequent to be overlooked—and the spills continue up to the publication date of this book.

One year Dave had received a request from Union Chemicals to give them a record for an entire year that the company's sludges had been disposed of at the Sterling site. Although the company had promptly paid all the bills sent to them, officials had no record for themselves of what Dave had disposed of for them, and the DEC had requested that information.

After Dave had taken his information concerning the DEC and Union Oil to the local papers, he had been cut off from any more business. Union Chemical couldn't dump Dave immediately, but they had quickly begun to ease out of the relationship. A total of $300,000 a year was hard to immediately dump, and the company had had a contract with Brown. And without Brown taking its waste, what would they have done with it?

After Dave's first newspaper ad had appeared, his business with Unocal Chemicals had dropped to $147,872.05 over the next eighteen months, or about one-third of what it would normally have been. The company had done this by eliminating overnight two-thirds of its trucking costs. Dave couldn't help but wonder that if this was so easy to do, why hadn't the company done it years before.

The second way the company had eliminated costs was that it no longer used the Sterling Special Waste Site. Of the $147,872.05, none of it had been for disposal fees at the Sterling site. It had all been for trucking and transferring within the plant. Frequently, while servicing the plant, Dave would notice the vacuum trucks of his competitors at the plant. But those trucks had never come to the Sterling site.

On page nine of Unocal Chemicals NPDES permit, it stated, "Oily waste skimmed from effluent basins is no longer sent to the Sterling disposal site . . . Unocal now burns all waste oil in one of its boilers under a state air quality permit." Also on page nine of the permit was the statement, "The Unocal Plant has an excellent history of compliance." There had been no mention of the 1974 incident [See Appendix BBB].

Page seven of the permit described what Unocal Chemicals had been doing with the sludge. "The GES and PES generate approximately 50 and 186 gpd (gallons per day) of sludge, respectively. (Estimates are based on the assumption that the annual quantities of generated sludge are generated on a daily basis.) The sludge is sent to a sludge dewatering system (drying beds), which results in 23 gpd of dried sludge. The dried sludge will be sent to a landfill or metal reclaimer. The liquid from the dewatering system is recycled to the GES and PES 144 and 168 gpd respectively.

"The dewatering system shown in the application is not yet fully functional. The Kenai Borough landfill has refused to accept Unocal's sludge in the past because of its high water content. The company found that centrifuging and belt processing were not effective in removing the water. They frequently found that freeze-drying is promising. To date, they have pumped the sludge into gravel pits at the end of the summer, then allowed the liquid to evaporate, or freeze or sublime in cold weather. Remaining liquid drains into a sump, which is pumped out. The residual sludge is granular and nontoxic. The company will again ask the borough about sludge disposal this year."

Dave accepted the possibility that the plant had begun skimming the oil and burning it. The company had placed some oil storage tanks on site, and he was aware the company had started skimming its pools.

But Dave balked at the sludges. The sludges had been so thick and gooey that after pumping the ponds dry of water, the vacuum truck drivers had had to get down on the pond bottoms with a long-handled squeegee to corral the remaining sludge over into the corner where they could suck it up with their vacuum hose. There had been tremendous quantities of this sludge and Union Chemicals could not operate the plant without periodically getting rid of it. The date of the public notice on the NPDES permit renewal had been July 26, 1988. The company had brought no sludges to the Sterling site after October 15, 1984. That was four years.

Dave still couldn't help but wonder if the sludge treatment process described in the permit were so simple, why had they waited so many years to discover it? While the permit had stated that the oily waste used to be taken to the Sterling site, why hadn't it stated that the sludges used to be taken there, too, and what had happened to the sludges since 1984?

Unocal Chemicals has a ten-inch outfall pipe which extends well out into Cook Inlet. This pipe branches off into two eight-inch pipes. An eight or ten-inch pipe will handle an awful lot of debris or sludge, no matter what kind it is. No accurate records will ever exist until the people of the State of Alaska start counting the dead fish, and until the clam and shrimp beds are contaminated beyond repair, or until all the crabs are gone and the inlet is posted with danger signs from Kenai to Seldovia.

No one except Unocal Chemicals will ever know what or how much is actually going through that ten-inch outfall pipe. The people of the state of Alaska must trust the company's integrity. Since the 1974 incident, the next two managers of the plant had been fully integrated into the community, building respect among the area's residents. As the area's largest employers, with respected managers, it would have been any accuser who would lose respect and standing in the community, not the accused.

Maybe it's not integrity, however. Maybe it's just a necessary watchdog. What is needed is a boat capable of taking samples from the bottom of the inlet, from a point above Kenai, all the way down the inlet. Considering the richness of the marine life in Cook Inlet, and how much the local economy depends upon it, both in the form of commercial fishing and tourism, it's hard to believe the state doesn't have the money necessary to ensure the inlet isn't being poisoned.

There are other industries still discharging their waste into Cook Inlet. There are fifteen offshore drilling rigs discharging everything—including drilling muds and oil and chemicals—into the inlet. The Tesoro refinery, the Chevron Refinery, the Phillips Liquefied Natural Gas Plant all discharge into the inlet. Oily bilge water is dumped into the inlet. The city of Anchorage dumps its sewage from its Point Woronzof Sewage Plant into the inlet. The plant only provides primary treatment of the sewage, and EPA waived the requirement of secondary treatment because of the cost. According to a newspaper article, "Anchorage dumps its sewage in Cook Inlet after passing through a primary treatment process at the city's Point Woronzof facility. City officials maintain further processing is unnecessary because the inlet's strong tides flush the partially treated sewage out to sea."

That has been the battle cry of all advocates of dumping into Cook Inlet—"It just gets flushed out to sea, so why not dump it?" That used to be the battle cry of industrialists in Puget Sound, too.

The city of Kenai, which had a good record for its sewage treatment, has had to post signs along its beach warning of the possibility of sewage contamination. The city had been dumping secondary-treated sewage, which is ninety percent free of contamination, into the inlet—leaving ten percent for the inlet to handle. That will change beginning in 1991, when the city will be required to chlorinate and then dechlorinate its waste before discharging.

The combined effect of all the polluters has had some environmentalists and fish and game officials worried for more than a decade. Yet there has been no action taken. No one has really scientifically considered the total amount of the discharges. When EPA reviews a permit, only that permit is considered, without considering what all the discharges might be doing.

In 1977, the State Fish and Game Department's official review of EPA's Environmental Impact Appraisal of then Collier's NPDES permit had several concerns, including:

"The combined effects of the Collier plus various other industrial effluents in the Kenai area; apparent adverse effects on fish migration in the Kenai area and potential effects on spawning; potential accumulation of contaminants in shellfish; and inadequacy of present monitoring programs to detect many potential adverse effects on the fisheries of the area."

The discussion had continued:

"It is stated that biological studies indicate no significant harmful effects have occurred to plankton in the area. Such study results are to be expected as a result of even minimal tidal flushing of the area and replenishment of plankton from other areas of the Inlet. Thus, even though significant mortalities may be occurring over a long period, replenishment of the area by other Cook Inlet stock will preclude detection of significant mortalities in the Nikiski area." In other words, Dave mused to himself after reading the paragraph, the kill is taking place, you just can't tell it by studying it right there.

The review concluded: "The discharge from the Collier Plant is a potentially serious problem which has not been adequately investigated or evaluated. Additional discharges from other plants in the area could compound the problem and result in damage to fish and game resources south of Nikiski. Further industrialization of the Nikiski area should not be pursued until the effects of industrial effluents on local fish migrations and downstream shellfish populations are studied in more detail. Due to lack of positive evidence that Collier effluents are adversely affecting fish migrations, the ADF&G considers it reasonable to approve Collier's permit request for modification of its pollutant discharge elimination system, but only on the condition that positive and immediate steps are taken to study and identify potential effects of local effluents on fish migrations in the Kenai area and shellfish populations in the Clam Gulch area.

"The EPA appraisal procedures should be broadened to consider the combined effects of all industrial effluents in areas of concern. Environmental effects of effluents cannot be adequately assessed by consideration of effluents produced by individual industrial activities."

By 1990, none of the above suggestions had taken place. A newspaper article written by Polly Crawford and published on August 8, 1988, in *The Peninsula Clarion,* had detailed the continuing problem of pollution and Unocal Chemicals.

"Disputed reports that Unocal Chemical is rated high enough to make the Superfund cleanup list, controversy over proposed air emission standards and questions over just what exactly is at the plant and where it goes all boils down to one main question in Nikiski: is people's health being affected?

"Many local residents believe it is and have complained to the Alaska Department of Environmental Conservation about such problems as asthma and breathing the irritating smell of ammonia.

"Unocal management says that not only are its emissions within federal and state standards, its current emissions of urea and ammonia are not harmful—with the ammonia levels around the plant often lower than a home kitty litter box.

"The company has been reporting its emissions and its spills to the DEC and the United States Environmental Protection Agency for many years.

"There are several reasons why the controversy is coming to a head at this time. First, the final Superfund report completed by Tryck, Nyman and Hayes entitled, "Suspected Uncontrolled Hazardous Waste Site Inspections," prepared for the DEC in conjunction with the EPA's investigation into potential Superfund sites—for which a special federal fund has been established—has just recently become available.

"Second, the company's National Pollutant Discharge Elimination System (NPDES) permit is up for renewal—the permit which outlines what the company can put into Cook Inlet. Plus exactly what materials the company has at its plant has become a matter of public record as of July 1.

"Third, the company is on the DEC's list for the special appropriation cleanup, in which DEC officials will oversee the company's cleanup operations.

"And fourth, the DEC's proposal to increase the opacity standard on the urea prill tower, coupled with a decrease in urea grains in the air and in the amount of ammonia allowed into the air, has brought a cry of foul from both the area residents and the company.

"The fact that the groundwater beneath the plant has been contaminated by ammonia, sulfinol and arsenic that is seeping out of the bluff into Cook Inlet has been widely known.

"But some other facts have only recently become known as a result of a perusal through DEC records.

"Although the Superfund report was recently made available, the draft Hazardous Ranking System (HRS) score, which determines whether the site must receive priority attention by the EPA, was excluded from the report because the number was still a "draft," not yet approved by EPA in Washington, D.C.

"Carl Reller, the former DEC employee in charge of the investigation, and all Superfund investigations in the state until he quit in December, said the score as proposed by the contractor was 22.98.

"That score, well below the required 28.5 for the priority list, did not include the company's air emissions, and Reller said EPA told the DEC that air emissions would have to be included.

"Reller, an environmental engineer who now works for the Alaska Center for the Environment because he claims of frustration with DEC lack of enforcement, calculated the score with the air emissions himself. The necessary raw data, he said, is included on the back pages of the report, and the Federal Register provides the instructions for calculating the score, including air emissions. 'A sixth grader could do it,' he said.

"His calculation brings the score up to 32.59. A site, however, doesn't get on the National Priorities List until the report and draft score are reviewed by a company called the Mitre Corporation, and officially approved by the agency. There are approximately 1,000 sites on the list throughout the country.

"But Unocal management disputes the use of air emissions as part of the score. Keith Laurie, in charge of permits for Unocal, said both the Superfund report contractor and DEC agree air emissions from an operating plant don't count toward a Superfund ranking—particularly when those emissions are within state and federal guidelines. He said the intent of air emissions in the ranking system is to account for measurement of volatile organics and other pollutants seeping up from the ground.

"The main focus of attention among residents in Nikiski has been the quality of the air. Marie Becker, whose 10-year-old daughter became afflicted with asthma, puts the blame squarely on the poor air quality from the refineries in general, Unocal in particular. Her house, she said, is four miles from the plant—upwind according to the prevailing winds.

"Her daughter's condition has become so bad, she said, she is often hospitalized. Yet away from home, she claims, the problem clears up.

"Becker's articulate expression of her concern has sparked others to link their problems to the air. Plus DEC has received periodic complaints over the past few years about cars being covered with white dust, foam on Bernice Lake, a strong odor of ammonia, and etching of auto glass surfaces.

"Ammonia is the primary pollutant which comes out of Unocal's stacks daily, and only a few states have standards for the amount of ammonia that can be put into the air. It is listed as 'an extremely hazardous substance,' in Unocal's report to the EPA, but Laurie explained it is only hazardous in extremely high doses. It is found in the air routinely and in most homes, particularly in homes with wet diapers or kitty litter. Anytime the odor of ammonia can be smelled it is far higher than the new proposed state standard, Laurie said.

"New CERCLA (Comprehensive Environmental Response Compensation and Liability Act) reporting requirements reveal that Unocal last year released 30,200,000 pounds of ammonia into the air, and will probably again this year. The ammonia is measured by knowing how much the gas weighed when it was a liquid before it was released. It boils at extremely low temperatures, so evaporates when it comes in contact with the air.

"While that averages out to about 40 tons per day, there are occasional spills that occur through the year. One such spill occurred on July 2, according to a required notification to DEC, and released 20,000 pounds into the air. Bill White, manager at Unocal, called the event a 'controlled upset,' in which, because of equipment problems, the ammonia was manually vented into the air.

"Anytime such events occur that release more than 5,000 pounds into the air, they must be reported. If more than 100 pounds of ammonia are spilled on the ground, it must be reported.

"Other spills, according to DEC records, have been: 6,000 pounds on Feb. 18; 10,000 pounds on Dec. 30, 1987; 15,000 pounds on Oct. 2, 1987; 16,000 pounds on June 3, 1987; 40,000 pounds on July 18, 1986; and 80,000 pounds on April 13, 1985.

"The company has three ammonia monitors, one at the southeast corner of the plant, one on the northwest corner, and one northeast of the plant across the highway on Miller Loop. It is required to report when the monitors show more than 25 parts per million of ammonia for more than 30 minutes.

"During the July 18, 1986, spill, 50 ppm were registered at one monitor. While most days don't register enough ammonia to report, there are days when normal operations cause the monitors to go about 25 ppm. For example, on Dec. 16, 1986, the northwest monitor registered 50 ppm, while no spill was reported that day.

"A chart distributed by the DEC puts the parts per million into perspective. More than 500 ppm causes immediate danger to life and health. More than 100 ppm causes irritation of the respiratory tract. The National Institute of Occupational Safety and Health recommends 50 ppm as the five-minute exposure limit. The American Conference of Governmental Industry Hygienists recommends 35 ppm as the 15-minute occupational exposure limit. The Navy's limit for continuous exposure in submarines is 25 ppm, and the complaint threshold among exposed workers generally is in the 20 to 25 ppm range.

"DEC is proposing 20 ppm for the 15-minute Alaska Ambient Air Quality Standard, and .5 ppm as the eight-hour standard.

"Most states, including California, where Unocal's sister plant is located, have no standards for ammonia in the air. But seven do, according to the DEC. Florida, Connecticut and South Dakota all have .5 ppm as their eight-hour concentration. Nevada has .596, and Virginia, Wyoming and Massachusetts standards are similar, based on different time periods.

"While both Laurie and White agree Unocal's ammonia emissions are probably greater than they should be, and that the company is working to reduce those emissions, .5 ppm off site is so low, they're not sure there's equipment available to monitor it. White added that there is 100 years of health data to back the claim that small amounts—larger than .5 ppm—cause no health problem.

"The proposal also caught the company officials off guard. While DEC has mentioned the possibility of setting a standard a year ago, Unocal officials were never contacted for their input into what the standard should be.

"Last year, White said, Unocal spent $8.5 million updating an ammonia plant and reduced the company's emissions from 50 tons per day to 40 tons per day. One of the urea plants is a major source of ammonia, and the firm is now undergoing a major revision on how it works that plant, White said, to bring down the emissions.

"He added that the ammonia which does rise into the air has the potential for reacting with the sulfur and nitrous dioxide emissions of other plants to form a natural fertilizer ammonium sulfate or ammonium nitrate—also formed by lightning.

"Ammonia is just one emission. The other major emission is urea dust, which, Laurie claims, when breathed is merely passed through the body.

"For several years, the company has been in violation of the opacity standard—how much light is passed through—on its plumes from the prill tower, where urea pellets are made.

"Based on a consent agreement with the DEC the company is currently installing equipment designed to get rid of the smaller dust particles—which make the plume look thicker —by revising the prilling process. The design is brand-new, developed by Laurie.

"Then the DEC, last week, announced the adopted standard would include a cap of .035 urea parts per million—the amount cited by Unocal as the proximate equivalent to 40 percent opacity. The DEC also raised opacity to 40 percent, from 20 percent. The previous grain cap had been .1 ppm.

"But the new process, while reducing opacity, may not reduce the actual grains. The prill tower has been averaging .034 ppm, but they cannot guarantee it won't go above that, and Laurie and White believe the unexpected standard was the result of DEC's desire to placate Nikiski residents because of past problems in which DEC was perceived as siding with industry.

"There are other substances at the plant that could be considered more harmful, and which are listed in the company's CERCLA report.

"A total of 18,000 pounds per year of 1,1,1 trichloroethylene, a degreaser found in spray cans available in auto parts stores, is evaporated at the plant. An additional 2,500 pounds is included as part of the waste stream into Cook Inlet.

"The compound is considered a carcinogen.

"Laurie said that while there are a few drums of the substance used, most of it is used in the form of spray cans, and there is no way to account for where it goes.

"While 1,1,1 trichloroethylene is considered a hazardous waste, according to Dave DiTraglia with DEC, if the waste is evaporated—unless on purpose—there's no regulation accounting for it.

"The 2,500 pounds through the year into the waste system is so small, Laurie said, it wouldn't be detectable.

"The presence of 1,1,1 trichloroethylene, a chlorinated solvent, is tested in waste oil by finding the amount of halogens—chlorine, bromines, fluorines. Halogens have been found in the company's waste products and waste oil. Usually, the amount is from 300 to 900 ppm, although one sample, in the fall of 1986, was 1,714 ppm of halogens.

"While anything more than 1,000 ppm of halogens makes the waste a hazardous waste, a loophole in EPA stands allows the waste to be burned in an industrial boiler as a fuel, Laurie said, because no standards have been set for industrial boilers. When burned, the chlorinated solvents and halogens form hydrochloric acid and chlorine gas.

"Bill MacClarence, DEC air quality engineer, was unaware of the 1,714 ppm report, and unaware of the EPA loophole. He said he was pursuing it, and Laurie said he suspected that future permits will probably require closer scrutiny of the amount of halogens in the waste oil and oxozolidone, a harmless waste product from the urea process.

"Another hazardous product burned in the industrial boiler is arsenic, also contained in the oxozolidone. Tests were done to show the arsenic emission at the stack is far below what is required—mostly because arsenic doesn't burn. White said that once a year the inside of the boiler is scraped and the soot and arsenic is barreled up and sent to Oregon to a hazardous waste dump.

"One incident involved a spill of 710 gallons of UF-85, urea formaldehyde. The spill was cleaned up and put into the company's effluent ponds, and at least 346 gallons were discharged into Cook Inlet. White said the company believed it was part of the NPDES permit and the required Best Management Practices plan. An employee called in the discharge, and EPA disagreed with Unocal's interpretation of the permit and called a halt to the practice.

"But the EPA does allow the company to discharge other things into Cook Inlet, and is currently revising the permit with about the same amounts of discharges, but with improved Best Management Practices.

"According to Kerrie Schurr, with the EPA, the biggest discharge is ammonia—1,952 pounds per day average, with a daily maximum of 3,863 pounds.

"Discharge of organic nitrogen from the urea plants is 2,850 pounds per day for a daily maximum of 5,330 pounds.

"There is also oil and grease allowed in the discharge—150 pounds per day or a daily maximum of 300 pounds. Schurr said the new permit will probably require toxicity testing of marine organisms subjected to the effluent. The public has until Aug. 25 to comment on the draft permit, which can be viewed in EPA's Anchorage office. The permit review period was advertised in the Anchorage daily papers (Anchorage is located 150 mountainous road miles from Kenai).

"While commenting that almost all of the information being used against the company is the result of Unocal abiding by reporting requirement—in other words, being a law-abiding company.

"Its own ground water problem was found back in the 1970's as a result of the company's soil sampling to put in another plant. The contamination was then reported to the state.

"The company is currently cooperating with the DEC in conducting a groundwater characterization to determine the flow and whether the lower confined aquifer has an impermeable clay layer protecting it from the currently contaminated upper aquifer.

"Then it plans to work with the DEC in addressing whether the seepage into Cook Inlet should be cleaned up, or whether the seepage is in reality not significantly more than the company is permitted to discharge into Cook Inlet.

"The DEC also plans to determine the cause of the acres of dead trees across the river from the plant. Laurie said the company believes the dead trees were the result of an early accident in which ammonia carbamate was released, stressing the trees and making them more susceptible to beetle infestation.

"He added ammonia carbamate release is no longer possible."

14

POLLUTION EVERYWHERE

By 1987, oil companies were getting nervous. Oil had brought misery to thousands. Oil brought repossessions by the thousands. Oil brought foreclosures by the thousands. Oil brought bankruptcies by the thousands. Oil brought banks closed and credit unions and savings and loans closed. Oil brought schools closed. Oil brought unemployment checks and welfare doles by the thousands. Oil caused thousands of Alaskans to lose everything that they had spent an entire lifetime gaining. Oil picked Alaska up and then smashed it into the ground, and, along the way, oil corrupted politicians and bureaucrats, poisoned Alaska's land, water and some of our people. Oil brought bumper stickers to many Alaskan cars and pick-up trucks that said, "Please, Lord, give me just one more boom. I promise not to piss it away next time."

But it wasn't only financial misery oil had now brought. It had brought even more serious consequences: pollution, illness and death. More and more people began to speak out against careless and intentional pollution; so more oil companies were being investigated. Also, by this time, the DEC had had a complete change of staffing in its Soldotna office, although the man in charge of the Anchorage office, Bill Lamoreaux, remained glued to his position.

Amoco does very little business on the Kenai Peninsula. It operates four offshore drilling platforms in Cook Inlet—the Bruce, the Anna, the Baker and the Dillon. It also has an onshore delivery facility or receiving station, at East Forelands, in Nikiski.

Dave was surprised when, at his new home in Seattle, he read the newspaper article by Polly Crawford, printed in *The Peninsula Clarion.*

"The Alaska Department of Environmental Conservation is investigating allegations that hazardous or other types of waste may have been dumped at Amoco's East Forelands Delivery Facility in Nikiski . . .

"According to Lester Leatherberry, manager of the Soldotna DEC office, no proof of the allegations has been found, although the data from sophisticated electronic equipment has not yet been analyzed.

"The investigation has been under way since July, when the DEC office received tips from people concerned that inappropriate dumping may have occurred, Leatherberry said. He said the area is the site of an old oily water separator pit."

Dave had to chuckle when he read the next part, which was an indication of how paranoid the oil companies had become.

"*The Clarion* learned of the investigation last week and obtained some internal Amoco memos. A memo dated July 30 states that Amoco is in the process of an ongoing internal investigation regarding the East Forelands Delivery Facility, and that all personnel were to be 'instructed that any communication with the Department of Environmental Conservation or with any state or federal agency is to be conducted only by Cheryl Winkler. All East Forelands records are to remain preserved and are under no circumstances are they to be destroyed or altered.' (The confusing double use of the very 'are' is an accurate quotation.)

"The memo also states, 'Further, should the press contact any Amoco employee regarding an Amoco issue, you are requested to direct the contact to me.' The memo is signed by W.G. Smith."

The investigation turned up no illegal dumping, and Dave was not surprised. While he couldn't vouch for what happened in the area fifteen years ago, he did quite a bit of work for Amoco. He hauled drilling mud for Amoco from the Nikiski dock to the Sterling site. He hauled water, he repaired heat exchangers, he rented them an acid pump, and he pumped out tanks for them. He transferred oil for them and skimmed pond oil.

But Dave had been most impressed with Amoco's cleanup job in 1983 of an outdoor storage pit that had existed for many years. The job was run by F.R. Peanut Melott with his crew of roustabouts. Dave furnished vacuum trucks and truck drivers. The pit contained mostly oily water, and, as Dave pumped it, Peanut and his crew used fire hoses to wash down the dirt. Dave contends that if ever any oil company on the Kenai did a good job of pit cleaning, it was Amoco that year. Dave remarked to Betty after the cleaning, "The earth under that storage pit is so clean I'd spread my sleeping bag on the ground and spend the night sleeping under the stars."

At the time, Dave was glad to see the DEC finally taking an interest in the oil companies; however, after seeing How Mr. Smith instructed his employees not to talk to DEC officials, but to route all communication through Cheryl Winkler, he couldn't help but wonder if the DEC was employing the same kind of sneaky tactics used against him.

On at least two occasions Dave discovered evidence that John Dough climbed over the chain link fence at the Sterling Special Waste Site, without permission and without a warrant. People from Ecology and Environment in Seattle, Washington, doing some testing at the Sterling site for EPA and the DEC, also climbed over the fence and locked gate. Dave was supposed to meet them to let them in, and he was on time. However, they were early, so they just climbed over the fence. Dave's temper flared, and he threw them out for illegal trespassing. Dave was then cited by a United States Marshal for interfering with an EPA investigation.

Dave believes DEC officials' actions hurt them in the long run. DEC officials investigated illegal dumping of raw sewage on the ground in the woods in Nikiski. DEC had the guilty party, but lost the case in court, because they got their evidence by illegal trespassing without a search warrant. The sewage never did get cleaned up. It lay on the ground for years.

Tesoro is another refinery on the peninsula, one fraught with environmental problems almost to the degree of Unocal Chemicals.

Dave had dealt with the refinery on numerous occasions, and found the personnel pleasant to deal with. The managers have been active members of the community, building up the plant's respectability.

Dave Brown did some vacuum trucking for Tesoro. He had done some steam thawing, and rented power washers to them. He did boiler cleaning and sand blasting, and, on several occasions, sent a power wash truck and men out to do cleaning. He cleaned heat exchangers for Tesoro using his fifteen thousand P.S.I. hydroblaster, and cleaned a big tank that had been damaged by fire.

In early 1982, after Dave had invested $127,000 in his little tank farm and waste oil cleaning equipment, he made arrangements with Tesoro to collect, at no cost to himself, all of the waste crude oil that he could pump out of the refinery's outdoor storage pit which contained a lot of A.P.I. Separator sludge. The outdoor pit was located behind the trees in back of the refinery. It was unlined and not fenced in, but it was out of sight. It was probably a little less than an acre in size.

Tesoro needed to get rid of the oil and clean up the mess. The existence of the pit had been observed by DEC officials.

Of course, EPA had reclassified A.P.I. separator sludge as a hazardous waste, and Dave could no longer use it. The pit sat for awhile.

Then Tesoro underwent a huge expansion, building

many new tanks and constructing gravel dikes around the pads on which they erected the tanks. Dave didn't participate in the building of those dikes, but heard that some of the remaining waste crude was mixed with the gravel in the dikes to get better compaction of the gravel.

A Kenai contractor showed Dave some photos of waste oil being land farmed into the ground with a dozer and vacuum truck, allegedly behind Tesoro. The oil was spread on the ground by the truck and then plowed in by the dozer.

Dave didn't know if that was the incident which eventually caused a furor at Tesoro, but something did—the result of public awareness of pollution which has made industry more conscious of what it is dumping into the ground, and how it looks to the public.

A newspaper article written by Ronnie Chappell, published on July 24, 1987, in the *Anchorage Daily News,* reads, "A department head at Tesoro's Nikiski refinery has been suspended without pay for ordering the illegal burial of an undetermined amount of oily waste in a trench behind the refinery in May 1985.

"The state has begun an investigation of the illegal burial. Company officials say they didn't learn of the trench or its contents until earlier this month, when a department employee mentioned it during an exit interview . . .

"Necessary (Refinery manager Mark Necessary) said he interviewed other workers at the refinery to confirm the story, and then immediately contacted state officials. The department head found to be responsible for ordering the disposal was removed from his position and suspended without pay . . ."

But that was just the beginning of Tesoro's problems. By the end of that year, Tesoro had discovered a huge contamination of its own groundwater, with as much as several feet of diesel sitting on top of the groundwater. The contamination was caused by leaks in underground pipes, and testing of the pipe system caused more leaks in the aging pipes. Tesoro had to hire an environmental engineering firm to install several wells to try to pump the diesel out of the water. The company kept discovering more and more diesel, concluding the spill was probably more than 770,000 gallons, creating an oil plume in the ground water three feet thick.

Tesoro's sister plant near Fairbanks, Mapco, has had similar problems, discovering a spill of at least 250,000 gallons of oil in the groundwater under the plant.

Tesoro's problems continued. The A.P.I. separator sludge pits were disclosed in August of 1988, and termed "hazardous waste pits" in an article written by Polly Crawford in *The Peninsula Clarion.* The public disclosure of the pits was the result of the company's attempt to close out the pits by burying them and putting on a cap, despite the fact that the pits are unlined. The proposed closure plan to the EPA required public notice.

Despite its environmental problems, Dave considered Tesoro to be one of the better companies in the oil industry, in that the company at least dealt fairly with individuals, including Dave. Dave was in a bind once, in need of payment quickly. Tesoro responded. Dave wrote the following thank-you letter to Glenn Jackson, the Tesoro plant manager in response. The Dave Brown referred to in a letter is no relation to the author:

"Dear Glenn:

"Recently we did a job for you cleaning Tank #21.

"Shortly after we finished the job we ended up in a little financial bind.

"I went to Dave Brown at Tesoro and explained my plight. Dave went to Mark Necessary. Mark called your Texas office, and a couple of days later my check arrived in Anchorage on your company's private jet, hand carried I believe by your company president.

"Glenn, this may have seemed to you a small thing, but to me it was a big thing. It came in time for me to avert a disaster, and actually the bill wasn't even due yet.

"It is gratifying to know that a company as large as yours can stop what they are doing, long enough to help out a company as small as ours.

"I guess what I am saying is that Tesoro cares about people, and that's rare in this day and age.

"Thank you kindly, Glenn, Mark, Dave and Tesoro.

Sincerely,
Dave Brown"

15

BREAKING LAWS

An outlaw is one who performs acts which are unlawful or illegal. According to Webster, an outlaw is one who deprives us of the benefit and protection of the law. No one individual nor corporation is above the law, regardless of political clout or economic contributions. Wilful unlawful disposal is a felony in many states, punishable by stiff fines, cleanup costs, and prison sentences. In Alaska, the state needs the oil industry's money so badly that it fails to enforce its own laws. It ignores the violators. It pretends that the pollution will never affect the people, and loses sight of the fact that children and grandchildren will inherit the state, its land and its pollution.

In the fall of 1958, the Alaska Oil and Gas Conservation Commission had been formed to issue the permits to drill for oil and gas. The permits cover drilling anywhere in the state, no matter who is the landowner. It also issues permits for offshore drilling out to the three-mile limit.

Other permits are required in addition to the drilling permit. If drilling is to take place in wetlands, a construction permit is required from the United States Army Corps of Engineers. A disposal permit for the waste drilling mud is required from the Alaska Department of Environmental Conservation unless the mud is being taken to a permitted site. If any mud whatsoever is going to be left on site, a disposal permit is required by law from DEC and applications for that permit must be made at least sixty days prior to any proposed discharge.

The Corps of Engineers' permit has nothing to do with the disposal of drilling mud, and the Corps is not authorized to permit the disposal of drilling mud.

The United States Fish and Wildlife Service or the United States Department of the Interior is not authorized to permit the disposal of drilling mud on federal lands or within the National Wildlife Refuges located within the state.

The Oil and Gas Conservation Commission has made two points quite clear to the oil industry in Alaska regarding drilling mud. Under Alaska Administrative Code Title 20, Chapter 25, Sectio 047, Pages 11 and 12, it states: "(A) Reserve pits. Special precautions must be taken to render impervious the confining surface of a reserve pit. (B) Upon termination or completion, suspension or abandonment operations, the operator shall proceed with diligence to dispose of or solidify in place all pumpable fluids, and shall leave the reserve pit in the condition that does not constitute a hazard to ground water."

The other point which the commission made clear to the oil industry officials is that one requirement in obtaining the drilling permit is that they also obtain all the other permits required by other state, federal and local agencies, and that failure to do so constitutes a violation of the drilling permit.

It was ridiculous to hear a DEC employee make the statement at Cannery Loop #2, "We don't know what the Corps of Engineers permit allowed." DEC officials know they are the only agency in the entire state with the power to issue drilling mud disposal permits on anyone's land, including wetlands. "The statements by DEC officials and Union employees appeared to me to be ploys to try to confuse the issue and misdirect the press and the public."

Slowly, Dave compiled a list of outlaw companies, including a few entities that were not oil companies, but who nevertheless drilled for oil or gas. Each company knew the lawful requirements for the necessary drilling mud disposal permit, yet they did not get permits in many instances. Each of these instances took place after the DEC solid waste regulations went into effect in July, 1973.

The names of some of the companies listed may cause confusion, because of partnerships and absorption. For example, British Petroleum Exploration Alaska Inc. absorbed Standard Alaska Production Co. as well as B.P. Alaska and Sohio. So, for the most part, British Petroleum Exploration Alaska Inc. is listed by the Alaska Oil and Gas Conservation Commission as the operator, even though, in many cases, the drilling and the non-permitted unlawful disposal were performed by the companies whom it absorbed.

Likewise, ARCO took over from Chevron on the Kenai Peninsula, and is now often listed as operator when the actual drilling and unpermitted disposal was carried out by Chevron.

Most of the wells drilled by ARCO on the North Slope were drilled with ARCO as the operator, but with Exxon as a fifty-fifty partner.

Sometimes a single reserve pit serves as temporary storage of drilling mud from two or more wells instead of one, but this does not happen very often on the Kenai Peninsula. If an oil company does not want to pay trucking or disposal costs, it is much cheaper to dig one pit per well and cover it up afterward.

To give an idea as to the quantities or volume of drilling mud involved per well, Union Oil on the Kenai Peninsula regularly dug reserve pits of four hundred feet by eighty feet by fifteen feet deep, and when a well was completed, the pit would be filled to a depth of eight to ten feet. Mud from a second well could never be added because the mud couldn't be stabilized enough to backfill over the top of it. Drilling mud has been disposed of illegally in about two thousand separate incidents.

Dave now knew the names, owners and locations of each one.

Outlaw Oil Companies Pipeline Owners

1. BRITISH PETROLEUM EXPLORATION ALASKA INC & associated companies which it has absorbed—B.P. ALASKA, SOHIO ALASKA, STANDARD ALASKA PRODUCTIONS CO.

Wells Drilled:	528	
Disposal Permits:	4	
Total Violations:	524	
% of pipeline ownership:		50.01

2. ARCO ALASKA INC., ATLANTIC RICHFIELD AND EXXON

Wells Drilled:	1,226	
(Most of these were drilled with Exxon as a 50/50 partner)		
Disposal Permits:	6	
Total Violations:	1,219	
(State failed to list Arco's Funny River permit)		
% of pipeline ownership		21.35

3. EXXON

Wells Drilled	15	
Disposal Permits:	1	
Total Violations	14	
% of pipeline ownership:		20.34

4. MOBIL OIL

Wells Drilled:	8	
Disposal Permits:	0	
Total Violations:	8	
% of pipeline ownership:		4.08

5. AMERADA HESS

Wells Drilled:	4	
Disposal Permits:	0	
Total Violations:	4	
% of pipeline ownership:		1.5

6. UNION OIL COMPANY OF CALIFORNIA

Wells Drilled:	39	
Disposal Permits:	0	
Total Violations	39	
% 0f pipeline ownership:		1.36

7. PHILLIPS PETROLEUM CO.

Wells Drilled;	2	
Disposal Permits:	0	
Total Violations:	2	
% of pipeline ownership:		1.36

8. AMOCO PRODUCTION CO.

Wells Drilled:	5
Disposal Permits:	0
Total Violations	5

9 MARATHON OIL CO.

Wells Drilled:	5
Disposal Permits:	1
Total Violations:	4

10. CHEVRON U.S.A. INCLUDING STANDARD OIL OF CALIFORNIA

Wells Drilled:	50
Disposal Permits:	2
Total Violations:	48

11. TEXACO INCLUDING TEXAS INTERNATIONAL PETROLEUM CORP.

Wells Drilled:	12
Disposal Permits:	1 (for 3 wells)
Total Violations:	9

12. CITIES SERVICE OIL & GAS

Wells Drilled:	7
Disposal Permits:	0
Total Violations:	7

13. SHELL OIL CO.

Wells Drilled:	11
Disposal Permits:	0
Total Violations:	11

14. GULF OIL COMPANY

Wells Drilled:	4
Disposal Permits:	0
Total Violations:	4

15. CONOCO INC.

Wells Drilled:	63
Disposal Permits:	0
Total Violations:	63

16. HUSKY OIL NPR OPERATIONS INC.

Wells Drilled:	27
Disposal Permits:	0
Total Violations:	27

17. SIMASCO PRODUCTION CO.

Wells Drilled:	6
Disposal Permits:	0
Total Violations:	6

18. AMAREX INC.

Wells Drilled:	3
Disposal Permits:	0
(Disposed in one instance at a permitted site)	
Total Violations:	2

19. LOUISIANA LAND AND EXPLORATION CO.

Wells Drilled:	3
Disposal Permits:	0
Total Violations:	3

20. VAUGHN PETROLEUM OF TEXAS, INC.

Wells Drilled:	2

	Disposal Permits:	0
	Total Violations:	2
21.	FOREST OIL CO.	
	Wells Drilled:	1
	Disposal Permits:	0
	Total Violations:	1
22.	INEXCO OIL CO.	
	Wells Drilled:	1
	Disposal Permits:	0
	Total Violations:	1
23.	COPPER VALLEY MACHINE WORKS	
	Wells Drilled:	1
	Disposal Permits:	0
	Total Violations:	1
24.	ALASKA ENERGY DEVELOPMENT INC.	
	Wells Drilled:	1
	Disposal Permits:	0
	Total Violations:	1
25.	H.G.&G. INC.	
	Wells Drilled:	1
	Disposal Permits:	0
	Total Violations:	1
26.	NORTH SLOPE BOROUGH	
	Wells Drilled:	13
	Disposal Permits:	1
	(To cover North Slope Gas Field)	7
	(Wells drilled years before permit was applied for.)	
	Total Violations:	8
27.	UNITED STATES NAVY	
	Wells Drilled:	2

Disposal Permits:	0
Total Violations:	2

28. UNITED STATES DEPARTMENT OF THE INTERIOR

Wells Drilled:	1
Disposal Permits:	0
Total Violations:	1

The United States Navy was involved for several years in oil exploration at what was formerly known as Naval Petroleum Reserve #4 or Pet 4. This area lies west of Prudhoe Bay. The easterly boundary of Pet 4 is about one hundred to perhaps one hundred twenty five miles west of Prudhoe. Eventually Naval Petroleum Reserve #4 became the National Petroleum Reserve, around the mid-1970s. At this point the Navy ceased to be involved and the United States Department of the Interior took over. It was also about this same time that the Alaska Department of Environmental Conservation Regulations went into effect. Apparently the United States Navy and the United States Department of the Interior hadn't gotten word yet regarding the DEC permitting requirements, because they drilled three wells and disposed of their drilling mud with no permit.

[In the Appendix DDD are other lists, several pages of permits issued or applied for which are for the disposal of drilling mud. The first group are for pit disposal of heavy mud. The last is for wastewater disposal injection. Heavy muds cannot be injected. Alaskan Crude applied for and received four mud disposal permits, but each one seems to be listed twice. Standard Alaska production has four permits listed, but again, it may be a case of reissuance of permits for the same two wells.]

Dave collected from the Alaska and Gas Conservation Commission information which listed a total of two thousand four hundred sixty-six wells. The first pages list wells that were drilled prior to the state solid waste regulations. The remaining ones constitute the illegal rape of the state of Alaska while the DEC and the state government were asleep—purposely or ignorantly—at the switch.

From July 1974 onward, approximately 19,746,721 feet of wells were drilled onshore in the state. Based on his own work with wells, estimating costs and volumes, Dave believes that for every foot of well drilled, between four and five barrels of drilling mud are used, wasted, and usually left in the pit.

In 1981, Dave had been charging one dollar and twenty-five cents per barrel for disposal, four dollars was the rate set by the Kenai Peninsula Borough based on borough research of disposal in the Lower Forty-eight. Trucking costs are figured separately, usually running seventy dollars to eighty dollars per hour. Dave never charged anyone more than three dollars and twenty-five cents per barrel for disposal, and never charged more than eighty dollars per hour for trucking.

Basing his calculations on four barrels per foot drilling, he believes at least 78,986,884 million barrels of drilling mud were covered up. Deducting from that amount the estimated number of barrels which were disposed of lawfully, Dave calculated that 70,048,434.4 barrels were unlawfully buried, mostly on the North Slope in the Prudhoe Bay and Kuparuk oil fields where conditions are more harsh and costs are higher than on the Kenai Peninsula.

Using the figure of four dollars and twenty five cents per barrel, the cost for disposal would have been $331,705,844.50. Trucking costs usually ran higher than disposal, but, to be conservative, Dave used the equal figure. He conservatively estimated, then, that the oil industry saved at least $663,411,689 by deliberately polluting the state. Of course, there was—and still is—no place to put the muds on the North Slope, so there would have to be the added construction expense of a fully lined, permitted site to accept all the waste from the North Slope. He believes that would bring the amount the oil industry saved to a conservative seven hundred million to one billion.

This estimate only includes onshore sites. While offshore disposal is still legal in Alaska, it isn't in other states. If Alaska caught up with the other states in its desire to protect its water, a six hundred million dollar to one billion dollar cost figure would be extremely conservative. It could reach one and a half billion dollars to resuscitate Alaska," Dave said angrily.

16

PROOF AND POLITICS

Dave Brown never found absolute proof that one or more DEC employees were being paid off to look away from the oil industry, but, in 1982, a DEC representative from the Anchorage office had been on the Kenai Peninsula and had spoken to both Dave and Ray.

During the conversation, he told Dave and Ray that he had received two or three anonymous phone calls suggesting that he should get smart and not take the drilling mud issue so seriously. According to this man, the caller suggested that if he'd lighten up a little on the drilling mud issue, he, too, could be the recipient of money and gifts as were other DEC employees.

This employee spoke to others about the phone calls, and generated a hand-written memo from DEC official Jerry Lane, dated June 4, 1982. The memo stated:

"No later than Tuesday (6/8/82) have a memo to me explaining in detail telephone threat you have received regarding your actions as a DEC employee. Also include and (sic) explanation of offers of money you have received, and statement made to you about other DEC employees taking gifts. Names and times are very important."

Rumors of the same kind of corruption continued for several years, coming from various sources. Even the FBI, for a brief period, considered initiating an investigation. The FBI is not familiar with drilling muds or environmental issues. They were concerned with corruption. A *Peninsula Clarion* clipping details it:

"The FBI is contemplating launching an investigation into the disposal of drilling mud on the Kenai Peninsula by Union Oil Company, according to an Anchorage agent.

" 'The Environmental Protection Agency has brought questions to our attention,' said Lawrence Nelson, agent in charge on the Alaska FBI. 'They would have no objection to our investigating the issue . . .'

"Nelson said the EPA normally would have primary jurisdiction over the investigation of drilling mud disposal. Since the EPA has indicated it does not have the resources to get involved, the FBI may take on the case, Nelson said.

"The FBI, however, is uncertain about what approach should be taken. Nelson said the agency has never investigated an environmental issue in Alaska.

" 'When you've not done something before, you don't enter into it willy nilly until you're sure that there's a reason to get into it.' Nelson said."

The FBI did not continue the investigation.

In 1986, Dave called on a prominent citizen and resident of Sterling. Ted Burns is a local businessman, contractor, gold miner, real estate developer and bank director. Dave had three conversations with him, and could think of no reason why the man would lie. In one of the conversations, Dave and Ted sat around a table in Ted's spacious home, with Ted's wife Rita seated close by, listening.

"You know, Dave," Ted said. "I was thinking about getting into the drilling mud business. I figured with the Sterling site closed, there'd be plenty of business. I went to the DEC with plans to land farm the mud, and the DEC guy told me I could get a permit. I thought I could make a bunch of money, because I have a lot of connections in the oil business.

"Boy, did I get a surprise! I got phone calls from Union, Chevron and ARCO saying that if I built the site they would not use it. Do you know why? They said they could dispose free of charge at their drill sites, with no interference from the DEC, so why pay for trucking and disposal."

"I'm not surprised," Dave interjected.

Ted continued. "I have a lot of information about payoffs being made." He named the oil company that he said was the source of the payoff money, and he named the employee of that oil company who allegedly was the "bag man." He added that the bag man's family and his own family were close friends, so if any of the information was ever leaked out, it would be embarrassing for both.

Ted went on to say, "I've seen the money because the 'bag man' drinks and talks too much. The money is in new $100 bills, and it comes from small contractors as the price they pay to do business with that particular oil company. This money goes into envelopes and is used to pay off the DEC officials so they will fail to enforce regulations."

Dave leaned forward and looked directly into Ted's eyes, "I need more information on how the situation actually works."

"It's very simple," Ted replied. "Let's say you furnish a D-8 cat to HYZ Oil with an operator for $130 an hour and you turn in a bill to so-and-so for one hundred hours. So-and-so then says to you, 'Where's that other fifty hours? You must have forgotten it on your bill,' so you redo your bill, raising it from the original $13,000 to a phony $19,500. HYZ pays your bill, you keep the $13,000 that you earned, but you bring the $6,500 that you did not earn and you give it to so-and-so in new $100 bills, sixty five of them. So-and-so gives the $6,500 to the 'bag man.' The 'bag man' keeps one half of the money, or $3,250, for the risk he takes, and he gives the other half to one man with the DEC. Supposedly, the DEC guy spreads that wealth around a bit, to his friends that are in on the pay off."

Dave wondered if he was being told the complete truth. He delved more.

"Is that all of it?"

Ted shook his head, "I have enough information on three men with HYZ Oil, all high-ranking employees, to cost all three their jobs." He named the three men and Dave nodded. He knew all three.

The story of how the money was distributed didn't shock Dave. It brought back memories of when he sold heavy equipment. Every welding machine sold to a certain international oil field contractor came overpriced. He had always felt the extra money was kicked back.

Ted also told Dave about one local contractor who was very unhappy and very concerned because he merely did what HYZ Oil Co. asked him to do, and he was well paid for it. However, that contractor was subsequently named as a defendant in three lawsuits because of the illegal dumping he did for the oil company. "You may be sure," Ted said, sarcastically, "he will get no aid or relief from HYZ Oil."

Dave had three different conversations with Ted: one in his home, one in a restaurant, and one over the phone. He had no reason to disbelieve him [See Appendix CCC].

Things did begin to change at the DEC office in Soldotna after Governor Cowper took office, early in 1987, and after Dave filed for bankruptcy and left the state. John Dough resigned and moved out of state, and another DEC agent was transferred. Lester Leatherberry was appointed as office manager and new field officers were employed.

Both Dough and the other DEC employee didn't leave the office with a clean slate, yet neither was it a dirty one. According to a May 12, 1987, article in *The Peninsula Clarion* written by Polly Crawford, "Both DEC employees have been the subject of various investigations because of their work with the DEC, but according to Bruce Erickson, coordinator in Anchorage of four district offices including Kenai's, nothing has been revealed by the investigations, nor did any of their activity relating to the Sterling site lead to their transfer or resignation."

On the borough level, the solid waste engineer in charge of overseeing the site took medical leave from the borough and was reportedly working in Florida in 1987. The borough position was then eliminated by the new incoming mayor, Don Gilman.

At least Dave could take comfort that he didn't go down alone.

On the state level, Governor Cowper appointed Dennis Kelso, formerly an official with the state's Department of Fish and Game, as Commissioner of the Department of Environmental Conservation. Kelso studied environmental affairs and soon became a clean-up advocate. Between Kelso and Cowper, one million dollars was freed up to begin cleaning up parts of the Kenai Peninsula. But one million dollars doesn't go very far, and much of the clean-up was directed toward obvious problems caused mostly by private individuals, particularly owners of leaky underground storage tanks, whose tanks were putting the volatile and water-soluble hydrocarbon benzene into local aquifers.

The internal organization of the DEC was also changed in 1987. The Kenai office was put directly beneath the Anchorage office, with Soldotna's Les Leatherberry answering directly to Anchorage's Bill Lamoreaux. Lamoreaux is the only one remaining on the scene in the state. Lamoreaux was the official whose decision not to hold a public hearing concerning Marathon's waste site in Nikiski caused so much furor.

Leatherberry resigned from the Soldotna office in 1990, and cited internal handling of enforcement issues as one of the reasons. He wouldn't elaborate on how he disagreed with the handling.

Meanwhile, the drilling mud pits dotting the peninsula remain where they are, daily leaching into the ground. The Sterling Special Waste Site remains where it was, too, despite the vocal cries of the Sterling residents, such as Walt and Elsa Pedersen and Leigh Smithen. The borough was left with the site, and had to decide how to close it out, and pay for the close-out.

17

MIXED PRIORITIES

To the unknown mother and small-town resident, politics in high places shouldn't affect the well water she drinks. Unfortunately, in the State of Alaska—and a lot of other oil-controlled states, it does.

She can cry, "It's criminal! It's wrong! It's my God-given and Constitutional right to have pure water to drink. It's in the Preamble to have the right to life, liberty, and the pursuit of happiness."

Clean water is synonymous with life. Without clean water to drink, there is no life. How, then, can it happen in a state like Alaska, where individual freedom is supposedly cherished above all else, that a person's water can be contaminated, and nothing happen to rectify the situation?

It's an easy answer: money and power. But how that money and power work is much more complex, and much more sinister.

The simple fact in Alaska is that politicians need money in order to gain power. The oil industry has money and that equates with power. Despite words that would have the general population believe the contrary, the candidate who has the most money in his campaign coffers is usually the winner. And the biggest campaign contributors in Alaska are the oil companies.

Therefore, if politicians want to be elected, they have to raise enough money to place their names before the public. Often, it doesn't seem to matter what that person stands for or believes, as long as the public perceives that person will take care of their pet projects. That makes politicians somewhat like used car salesmen. What they tell you you are going to get, and what you actually receive, are quite often not the same. These are unfortunate facts of life, nevertheless. And that concept usually hits the voters too late—after the water is contaminated.

In grade school, children learn that when you borrow something, you must give it back. In high school and college accounting, for every debit there must be a credit. Socially, even young people quickly learn that if a friend in a bar or restaurant buys a round of drinks or lunch, another round or lunch is expected in repayment. In business, when a company borrows from a bank, it must be paid back with interest. If it isn't paid back, the bank can get rather nasty.

Most politicians are no different. When they accept a contribution from someone, at some point, a payback is expected. Often, when the campaign contributions are from many small, individual contributors, the expectation is that the politician just do a decent job in representing them. But even for those small contributions, the givers expect to be able to have the ear of the politician when the time comes.

When the contributors are larger and the contributors are companies, businesses, or special interests of some sort, there is the expectation that the politician will support the kind of legislation favored by that contributor. On top of that, politicians who begin with the idea they're going to buck tradition, and stand up for what is right, regardless of whom their contributors are, soon find that the easy money which put them into office can just as easily get them out. For many, such money is a potent tonic.

In 1974, U.S. Senator Mike Gravel had approximately seventy thousand dollars worth of campaign debts paid off for him by the oil industry as a contribution because of work he had done on the industry's behalf in the past, and which he would continue doing. Almost half of that money was raised by the head of a company directly connected to the oil industry. In addition, thousands of dollars of his personal debts were paid by honoriums from various groups in the oil industry. His largest fee was $2,500 from Humble Oil, one of the major companies in the Alaska pipeline consortium.

On February 17, 1985, an Associated Press article by Paul Jenkins on the front page of the *Anchorage Daily News* was headlined "Sohio received help for permits" and "Sohio overcame scientific objections, legal questions to win Beaufort Sea permits."

The article stated: "With help from friendly politicians, Sohio steamrolled over scientific objections and legal questions to win permits needed to begin the first oil production in Alaska's Beaufort Sea.

"The company and its partners won a vital U.S. Army Corps of Engineers permit after the agency drastically and unexpectedly slashed its requirements for a causeway, reducing the need for breaches—gaps to allow fish to swim through—during construction."

At the end, the article stated: "Stevens [Senator Ted Stevens, Republican from Anchorage] received more than $35,000 in contributions from the project's partners, their subsidiaries, or employees and political action groups.

" 'I can tell you there is no connection,' he said.

"Young [Republican Congressman Don Young from Fort Yukon], received more than $10,000 from the partners, their employers or subsidiaries and political action committees.

"Murkowski [Republican Senator Frank Murkowski from Fairbanks] has received at least $15,000 in campaign contributions from the project partners, the subsidiaries, employees or political action committees.

"Senator Murkowski said he was not trying to influence the E.P.A. decision, instead he was just trying to do his job.

And he was angered by a question of whether the contributions influenced his involvement. He called it a 'cheap shot and shoddy journalism.' "

The peddling of influence over permits has continued to this day. Recent Alaskan decisions by the United States Army Corps of Engineers concerning the building of causeways in the Beaufort Sea have been overturned by the Pacific Northwest Regional Corps Office, with the stipulation that future decisions will have to be approved by the Corps' Washington, D.C. office. Of course, one can only speculate about what kind of influence Alaska's Senators and lone Representative in Washington, D.C. would have over Washington, D.C.-based decisions, when the decision-makers are only a quick phone call away and are part of the Washington bureaucratic clique which responds quickly to inquiries made by the politicians.

By the end of 1987, an oil spill in Cook Inlet threatened what was predicted to be the largest red salmon run in Cook Inlet's history. It did prove to be a huge run, but there were also tainted fish and fishing periods left vacant because of floating oil.

Then came the 1989 wreck of the Exxon Valdez, which spilled eleven million gallons of oil in pristine wilderness waters rich in marine wildlife. Among its first effects was that it proved costly to fishermen not only in Prince William Sound, but also in Cook Inlet and Kodiak as fishing for the season was halted. The bag of rags had riches hidden inside, however, when fishermen leased their boats to Exxon to help in the clean-up, and reaped large financial benefits. However, the long-term environmental damage and serious genetic effects will not be known for a long time, if ever, as it will be nearly impossible to connect fishing habitat or reproduction problems ten to twenty years from now with the 1989 spill.

During the horrifying weeks after the spill, when it became evident that any clean-up attempt was going to be a pittance against the enormity of the spill, and Exxon was condemned by nearly everyone, Representative Don Young continued to oppose oil spill legislation—even when his Alaskan colleagues, Stevens and Murkowski, supported it.

Even so, Young continues to be put into office by the Alaskan voters and has the support of much of the oil industry.

Federal legislation governing oil lease sales, the pipeline, and regulations such as allowing single-hulled tankers into pristine ports like Valdez, are vitally important to the oil industry; that is why the industry continues to finance the campaign of the current three representatives. But state politics can get even muddier. It's the state that regulates things like the dumping of drilling mud, or creates the hoops a company has to jump through in order to get a permit to drill.

Governor Bill Sheffield held office from 1982 to 1986, during a period of time when oil revenues were high. Where there is a lot of money, there are bound to be a lot of scandals. Among the scandals which plagued Sheffield was the North Slope Borough scandal, the Fairbanks office lease scandal, the VECO payroll deduction plan, the $2.7 million remodeling of the governor's mansion in Juneau, Sheffield's whirlwind fund-raising tour at oil industry expense. Many of Sheffield's heavy supporters were given cabinet posts.

The governor's mansion in Juneau is no palace. It is an old white house containing 12,900 square feet. Governor Sheffield remodeled it at a cost of 2.7 million, or $209 per square foot. At that price, a new mansion could have been built. Regardless, when Governor Steve Cowper took over in 1987, the roof leaked so badly that Cowper had to use pails on the floor to catch the water. And while money was pouring into the governor's mansion, the DEC laboratory was in such poor shape that wind and dust blew through the cracks in the building's walls.

In the North Slope Borough, there have been unceasing scandals in the headlines. The news started breaking regularly around February 1985, and hasn't stopped.

North Slope contractors—and everything in the North Slope is connected to oil—contributed heavily to many Alaskan legislators. Slope contractors also contributed $35,000 to the campaign of former Governor Sheffield. All of this may or may not have had anything to do with the fact that initially Sheffield was not interested in having the state study North Shore Borough finances. The headline on March 9, 1985, in the *Anchorage Daily News,* read: "Hush surrounds $500,000 house for slope official." On March 5, 1985, the headline blared "Sheffield received funds from North Slope contractors." On March 31, 1985, it read, "Slope firms contributed $250,000 to legislators." On May 16, 1985, the headline referred to Lieutenant Governor Steve McAlpine receiving $30,000 from lobbyist Lewis Dischner, and on July 4, 1985, the headline read, "State asks McAlpine for documents on $30,000 fee." McAlpine claimed the $30,000 from Dischner was a legal fee and not a political contribution and was cleared.

Lobbyist Dischner was not so lucky. Dischner was recently convicted, together with consultant Carl Mathisen, on federal charges of extortion, racketeering, and fraud in the longest, most expensive, criminal trial in the state of Alaska. They were accused of using positions as advisors to former North Slope Borough Mayor Eugene Brower to take millions of dollars in kickbacks and steer construction contracts to firms they secretly owned.

During the course of the trial of Dischner and Mathisen, allegations made by a witness, Thomas Gittins, cast Senator Al Adams from Kotzebue in a bad light. Gittins, an Anchorage contractor, testified that he paid Dischner nearly $300,000 in 1984, ten percent of his borough contracts, to represent him before the Brower administration, but also to protect him from the spiteful actions of Senator Al Adams. Gittins claimed to have paid Adams $772,668 over a two-and-a-half year period beginning in 1982, consisting of commissions and salary for developing business for Gittins' construction company. Gittins claimed that Adams produced a useless local hire list that he believed was sold to Enserch or Sohio in 1985 for, as he recalled, $500,000. Reference was also made to a letter demanding payment of $500,000 from Enserch to UIC Construction for "smoothing political obstacles" in the borough and "coordinating local hire." The letter also referred to UIC's assistance in obtaining permits for Standard Oil's Endicott project. Gittins' allegations concerning Senator Adams remain allegations.

Governor Sheffield had already made the headlines in early 1983 when he took what the newspapers described as a whirlwind after-the-fact fund raising tour at oil industry expense. According to the reports, he took a free ride to several cities in the Lower Forty-eight, including Dallas, Texas. Governor Sheffield had waged a two-million-dollar campaign in 1982. He spent a considerable amount of his own money on the campaign. Then he recouped more than three hundred thousand in 1983 as a result of that oil company-sponsored tour.

Governor Sheffield made the headlines again on April 20, 1985, when the *Anchorage Daily News* stated that those who contributed toward Sheffield's campaign received jobs in the Sheffield administration. "At least sixty-nine people who donated money to Sheffield's 1982 campaign ended up on Alaska's payroll or sitting on the numerous boards and commissions which advise state government." At least six of these contributors ended up in Sheffield's cabinet.

Attorney Norman Gorsuch gave $2,500, plus some contributions from his law firms and its partners. He became Attorney General.

Jack Griffin, an Anchorage bar owner, gave $1,675. He was appointed to the Alcohol Beverage Control Board.

The Fairbanks office lease scandal became a severe handicap to any further political aspirations of Sheffield. The state of Alaska needed office space in Fairbanks. The governor's good friend, labor leader Lenny Arsenault, had raised $92,000 for his 1983 campaign. According to the requirements of the lease, the only building in Fairbanks that met specifications was a building owned by Arsenault and his partners. Adding to the furor, on April 30, 1985, the newspapers stated that Sheffield gave lease data to Arsenault. According to the reports, he supplied no similar data to other bidders.

John Shively, the governor's Chief of Staff, left his post over the mess. He was found guilty of lying on the stand. Sheffield's memory failed him and charges against him were never proven.

The state of Alaska hired a former Watergate prosecutor to handle the Fairbank's lease probe, since Alaska's Attorney General at that time was appointed by the governor, and could not be expected to prosecute the man who hired him. Not to be outdone, Governor Sheffield hired a former Watergate attorney for his own defense—at attorney fees which were eventually reimbursed by the state.

On July 3, 1985, the *Anchorage Daily News* headlines read "Jury finds Sheffield unfit" and "Governor finds critical reports unbelievable." On July 4, 1985, the headline read "Lawmakers meet to decide on impeachment."

During the first few days of August, the newspaper carried stories stating that "the senators were going easy now with Sheffield," and that the governor wouldn't be impeached. On August 7, 1985, when his job appeared to be safe, he let the public know that he was "still a force to be reckoned with."

On August 18th, an *Anchorage Daily News* article carried a story titled, "Courage missing among state senators." The gist of the article was that no senator wanted to find himself in the position of having sent Governor Sheffield to the gallows if he was not going to be hung. To have to face the governor and deal with him daily for the remainder of his term—in the event he was not found guilty and impeached—could have proven to be extremely uncomfortable, and the senators were not willing to bear that potential discomfort.

The governor's race, being statewide, is going to attract more oil dollars than districtwide races, and, since the local districts are smaller, it is a little easier for a person concerned about environmental problems to get elected to the state senate and house of representatives. Still, the oil industry heavily backs those who will support its interests. A lot do. There were at least enough supporters of oil during the oil heyday—the first part of the 1980's. State budgets were bulging with excess money for buildings and employees, and the state was handing out money to local entities for huge and expensive palaces, such as a lavish performing arts center and four-lane roads which go nowhere. At the same time, budgets were sliced and regulations were relaxed in the Alaska Department of Environmental Conservation.

Alaska's political system has always been an embarrassment to the local residents, and, a few years back, there was an effort to clean up the system by passing a law governing campaign contributions. Basically, the new law stated that it would be illegal for any politician to receive a contribution in excess of one thousand dollars from any single source.

The passage of that law did not stop high-level campaign financing. A few individuals and companies figured out ways to contribute anyway.

One of those was VECO, a huge international oil field service company firmly entrenched in Alaska. After the law was passed, VECO put into effect a payroll deduction plan. By withholding funds from most of the employees on its payroll, VECO was able to come up with huge sums of money which it contributed to politicians. Veco claimed that since the money came from VECO employees instead of from the company, and since no single employee contributed more than the one thousand dollars legal limit, then the contributions had to be legal.

There were rumors that VECO's plan was apparently master-minded, at least in part, by former state senator Ed Dankworth, who left office in 1982, under indictment for two counts of using his office for personal gain. The charges were later dropped when the courts ruled that he was protected by legislative immunity.

VECO's employees subsequently contributed more than $100,000 into various Republican campaigns. State Senators Jan Faiks of Anchorage received $8,200, Don Bennett of Fairbanks received $8,250, Pappy Moss of Delta Junction received $8,220, and Paul Fischer of Soldotna and Richard Eliason of Fairbanks each received $8,200. The contributions were from VECO and a VECO subsidiary, Norcon, Inc.

Faiks, Bennett and Moss were also flown aboard a VECO-chartered plane to Prudhoe Bay, where they enjoyed a steak, lobster and champagne party in the VECO dining room.

In February, 1985, the Alaska Public Offices Commission ruled the VECO payroll deduction plan illegal and fined

VECO $72,660 plus $480 per day for as long as the dispute remains unsettled. VECO asked the court to lift the fine which the court did not do. Veco then filed suit against the Alaska Public Offices Commission, but lost.

Laws in Alaska are made, it seems, only to be broken by the lawmakers. Laws in Alaska are only to be rewritten when they become uncomfortable for the lawmakers and their friends.

Many legislators are like attorneys. They will listen to you briefly and then forget you unless you bring money, and lots of it. Large transfusions of cash seem to gain and hold the attention of these type of legislators. Perhaps this is why so many attorneys enter politics. In addition to the fat salaries, the campaign contributions, the speaking fees, the consulting fees, and, after retirement from politics, the lucrative lobbying fees are far greater than what most attorneys could ever earn in the normal practice of law.

A few years back, it came to light that several of Alaska's legislators had been accepting free airplane trips to the Lower Forty-eight from the oil industry. The legislators involved claimed they couldn't see anything wrong with it.

In the September 22, 1985, *Anchorage Daily News,* an article by reporter Richard Mauer carried the headline, "Laundered money uncovered." Illegal gifts tainted three Alaska campaigns.

A *Daily News* investigation uncovered the fact that Lieutenant Governor Steve McAlpine, Senator Jay Kerttula, and North Slope Borough Mayor Eugene Brower were implicated in election law violations.

The article stated that at least four contributors in three states let their names and bank accounts be used to launder $1,000 contributions by people who may have already given $1,000—the maximum allowed by law.

At least one of the McAlpine contributors had almost a total lack of knowledge of Alaskan politics. She did not know the governor's name, and she asked, "Is McAlpine the Lieutenant Governor?" Yet she had supposedly donated a $1,000 cashiers check to a man she didn't even know.

On August 20, 1985, in the *Anchorage Daily News* there appeared an article headlined, "Oil Industry leads Bush campaign contributions." The article insinuated that the oil industry controls the state's United States Senators and Representative, the state legislature, even one of the state's daily newspapers. VECO purchased the *Anchorage Times* in 1989. What else do they control?

They think they own the oil, and want everyone to believe they do. The oil companies and their political representatives fail to tell the public that the oil is owned by the public, and that the public is getting just a tiny portion of the profit while opening the state's gates to the rape of the land which will profit not the people of the state of Alaska, but the oil companies. The royalties from the oil which are flowing from this state have provided nearly all of the state's income, and numerous politicians who are also financed by oil, and the oil industry itself, have gone to great lengths to convince the Alaskan public that it should be grateful to the oil industry because it is the industry which is financing everything in the state. The oil companies have spent a lot of money on slick television commercials to try to get the public to believe they are doing the state a favor by developing the resources, and that the state should kowtow to the oil industry's wishes. To this day, anyone in Alaska opposed to opening development of the last untouched Arctic coast, within the Arctic National Wildlife Refuge (ANWR), is in danger of scorn and blacklisting by the oil companies. The Trans-Alaska Pipeline was rammed through Congress as the result of an embargo by Arab oil-producing nations, which resulted in a fake oil shortage. With Iraq's invasion of Kuwait, there is talk the event will push through development of ANWR.

Then when the oil is gone, the oil companies will walk out of the state with their pockets lined with oiled greenbacks, leaving behind oiled beaches, oiled soil and oiled groundwater.

18

BACK HOME 1990

As the plane circled on final landing, the sight of Anchorage's new visage—glass, steel and cement instead of the rugged wilderness of his memories—brought Dave Brown hurtling back to reality.

When Dave got off the airplane and began driving through Anchorage, remembrances of his former life—the good times—fluttered through his mind. The Chugach Mountains in summer still had snow icing on their peaks, and the fingertips of fir tree branches wore green nail polish.

"Damn," he muttered, "it feels good to be back home."

But an hour later, driving downtown, he suddenly noticed that hovering over the city was a purplish-brown cloud. Heading toward the West Chester Lagoon, he scanned sign after sign warning people about the pollution there. Leaving there he drove past strip mall after strip mall, still vacant, still a testimony to the crash of the oil prices in 1986. Later, he saw the flashy Loussac Library—a huge, castle-like building which housed the majority of the books in Anchorage. He saw the garish Performing Arts Center. "Jammed with oil money," he muttered. "Oil oozing from between the bricks, and greed dripping from the roofs and blinking from the gaudy lights of the Performing Arts Center which flashes to downtown visitors."

He remembered how such building had come about during the lustful and freewheeling 1980's in Alaska, when community competed against community to see which one could get the most money from the state. Useless yet extravagant projects were devised behind closed doors, and if representatives sent to the legislature couldn't bring home the state oil-slicked greenbacks, then those representative were no good. They had to be thrown out. Communities went as far as hiring their own lobbyists just to make sure they got the money to which the city fathers thought the community was entitled.

The legislature neared anarchy every year, as pet projects for a representative's city were held hostage for that representative's vote for another representative's pet project. And each election year, the votes went to the person who could bring home the most money, because the person who could bring home the most money was almost always the same person sleeping with oil. But that didn't matter. Oil was a necessary and very fruitful evil that the state welcomed.

Then came the crash. The price of oil dropped from $34 a barrel in 1980 to its low around August of 1986 of $9 a barrel.

The state of Alaska, which was dependent upon the oil industry for eighty-five percent of its revenue, suddenly found that revenue cut by approximately seventy-three percent.

The result was instant chaos.

Many people—in the state government, in the school systems, and in the oil industry—either lost their jobs or suffered severe pay cuts. The repercussions included loss of jobs in other sectors as the downward whirlpool into the drain began. People left the state by the thousands, and repossessions, foreclosures and bankruptcies kept the courts in a spin.

In Anchorage, Alaska Mutual Bank failed. Security National Bank failed. Alaska Bank of the North failed. United Bank Alaska failed. Alaska Continental Bank failed. Alliance Bank failed. Peninsula Savings and Loan failed. The old Alaska

Pacific Bank is now Key Bank. United States Senator Frank Murkowski was once the president of Alaska Bank of the North, and former Alaska Governor Bill Sheffield was on the board of directors of Alliance Bank when it failed.

Foreclosures quickly overtook new sales. People lost their jobs or had their pay cut and they could no longer afford the $1,200 per month mortgages. So many people were in the same boat that the market soon became glutted with homes for sale. Condo units plunged from $100,000 to $30,000 or less, overnight, and even at that price no one was buying. In the real estate multiple listing service for Anchorage, the Matanuska-Susitna Valley and the Kenai Peninsula, there used to be more than two thousand five hundred realtors. That number plunged to around eight hundred.

In 1984, there were three hundred eighty-five bankruptcies filed in Alaska. In 1985 there were four hundred thirty bankruptcies. In 1986 there were six hundred seventy-one bankruptcies filed. In 1987 bankruptcies catapulted to one thousand, one hundred and fourteen for the year, and in 1988 the number sprang to one thousand, four hundred and seven. Four thousand, four hundred and thirty-eight bankruptcies and still rising.

Schools closed down because of lack of students and lack of funds. Of course, when families moved out, so did the kids. The Kenai peninsula completed two new high schools in 1987, fed by the promise of continued growth. One of the schools was opened, because of the pressure from families who lived near the new school and didn't want their kids to continue to have long bus rides. The opening cut in half the population of Kenai Central High School, and curtailed its programs. The other school, Skyview, sat unused for two years. A new elementary school was also completed that year. It was supposed to alleviate crowding in the other two schools. Instead, the district closed down one of the other, older schools.

Numerous businesses had failed. Some had changed hands, but failure due to Alaska's downhill economic slide has been the reason in most cases for the sales. On the Kenai Peninsula, a fine old hotel, Merit Inn, had closed. Formerly known as

Sheffield House, it was at that time owned by Alaska's former governor, William J. Sheffield. In Soldotna, the International Inn and the Riverside House had closed.

The motels and hotels which are still open on the Kenai Peninsula received numerous cancellations of reservations due to the Prince William Sound oil spill. One fishing resort received $123,000 in cancellations.

The massive Exxon spill now covers not only the Prince William Sound, but it has reached Ressurrection Bay at Seward. It has entered the Cook Inlet and is now in the Shelikof Straits between Kodiak Island and the Alaska Peninsula, lapping at the shores of Katmai National Park. Among other solutions, Exxon proposes is to incinerate twenty tons of dead animals.

The oil industry pollution on the Kenai Peninsula, which Dave called to the public's attention in 1984, still exists. Although studies have been made, papers have been shuffled and a private consulting firm was hired to look into illegal disposal, little has been accomplished.

Early the morning after Dave arrived, he headed over to the Kenai Peninsula. There were plenty of people with hands out to shake: some happy to see him back, others smiling ironically, shaking their heads. They're probably thinking, he thought to himself, there's no fool like an old fool. He drove over to the Sterling Waste Site; it was closed out.

He checked on the multi-million dollar P.C.B. clean-up in the Swanson River Oilfield. It was in high gear, but not yet near completion. There had been no action taken on what Dave knew to be the many drilling mud pits scattered throughout the field. Getting out of his car, he trudged around the "No Trespassing" sign. The gate at the entrance to the oil field, just off the Swanson River Road, remained in place.

Later, he drove over to the Kenai Gas Fields. The oil sheen remained, and the drilling mud pits were still in place. The mud at Cannery Loop #2, which he had uncovered, was still there. He heard later that the DEC had concentrated on the private individuals with leaking underground storage tanks instead of the oil industry.

Greeting some of the residents, Dave quickly learned that most of the people he knew who had to live near the West Poppy Lane gravel pit still lived there, although their lawsuit is supposed to go to court in the near future. Mike Pelch, with 34,000 barrels of drilling mud still illegally buried in wetlands on his property, remains mired in his mud. He can't sell, or even give away, his land.

Dave heard persuasive rumors that Unocal Chemicals still legally discharges approximately one million gallons per day of gunk into the waters of the Cook Inlet, and large amounts of chemicals were still legally discharged into the ground.

Driving over to Cook Inlet the tide was low. Getting out of the car, Dave took off his shoes and stood on the beach alone. Dave saw for himself the black ooze which still dripped from the bluff into Cook Inlet. There still were daily dumpings of poison into Cook Inlet. He let his eyes roam over the pockets of brown-tinged polluted waters and pondered the future of his state. Even though the oil industry has polluted Alaska's North Slope . . . even though one company has spilled oil up and down the pipeline . . . even though the Valdez harbor is contaminated with hydrocarbons and oil . . . even though the Kenai Peninsula is dotted with illegal oil industry dumpsites . . . even though Exxon coated more than one thousand miles of Alaska's coastline with black crude oil, killing thousands of birds and animals—there *had* to be a future. The Alaskan public and politicians have to make a stand. Angrily he thought, they are all lulled to sleep by the money lining their pockets, and with another era of high-priced oil within sight, the slumber could extend into coma and death.

"If we let it," he muttered shaking his head in refutation.

"The oil industry will still play its poisonous lullaby. Fed by the acrid aroma of ammonia and the sight of filthy oil rigs silhouetted by the beauty of the Redoubt Volcano and the wilderness it represents, Dave Brown began to think out loud and write down his thoughts and new plans, some realistic and some fantasies, about what needs to be done to resuscitate Alaska.

"We need to stop writing laws and start enforcing them. We need regular inspections of drill sites, tankers, and pipeline camps. We need background checks on tanker personnel and double-bottomed tanker hulls, not only on new tankers, but on old tankers as well. Take the old tankers out of service one at a time, and return them to service only after the new double bottom has received a Coast Guard inspection and certification. We need heavy fines, clean-up costs, and jail sentences for the chief executive officers of the violating oil companies. We need more watch dog effort and responsibility from the United States Coast Guard, the Alaska Department of Environmental Conservation, the Alaska Department of Fish and Game and the Alaska State Troopers. We need to organize the people as pollution stoppers, similar to crime stoppers, whereby an oil company employee or an ordinary citizen can anonymously report incidents of illegal dumping, accidental pollution, or a potential hazard such as a tanker crew drinking. We need to give the E.P.A. and the United States Army Corps of Engineers strong enforcement powers in Alaska. They have no enforcement powers at present. They can only observe and evaluate.

"We are going to have to find a way to make lobbying a less attractive profession than it is. We are going to have to find a legal way to mercilessly squash any oil company or oil field service company that participates in any scheme for massive campaign contributions to politicians or political candidates, so that the oil industry can no longer control our elections, and so that the practice of selective enforcement will cease.

"We need to realize that willful pollution is an extremely serious crime and that the biggest criminals in Alaska are the ones who wear gray flannel suits or pinstripe suits, and who drive a Mercedes, a BMW, or a Jaguar, and who sip martinis at the Anchorage Petroleum Club. Anyone who willfully, illegally dumps his waste, risking pollution, is no better than a drug dealer and deserves no better treatment. Pollution causes cancer, birth defects, mental illness and other serious health problems, many of which can be fatal.

"We need to stop negotiating with the oil companies. If the offending companies fail to move their messes, then the state should do it for them and collect triple damages from the offending oil company.

"We need to make illegal disposal a felony instead of a misdemeanor, and we need to put a few chief executives behind bars now. We need to leave them behind bars without bail until their companies have cleaned up the messes which they know they have made. Give these executives plenty to eat while they are behind bars, but allow them to quench their thirst only with water from the wells which they have polluted. In other states, illegal dumping is a felony punished by stiff fines, cleanup costs and jail sentences. Why should Alaska be different?

"We need to have an attorney general who is elected by the people rather than appointed by the governor. It is difficult for an appointed attorney general to investigate a crime or scandal in which the governor who appointed him may be involved. It is difficult also for an appointed attorney general to investigate situations involving friends of the governor.

"If I were Alaska's governor, here is what I would do with the offending oil companies.

"First I would tell my attorney general to get a court order demanding that each and every oil company guilty of unlawful disposal submit within forty-five days a plan for the cleanup and proper disposal of its unlawfully disposed of drilling mud. That court order would also state that the offending oil companies would then have forty-five more days in which to gain approval or disapproval by the governor's environmental task force, which would be a group consisting of representatives from the DEC, from the Alaska Department of Fish and Game, from the Alaska Oil and Gas Conservation Commission, from the United States Corps of Engineers, from the United States Environmental Agency and from the United States Department of the Interior.

"This would give the oil companies a total of ninety days to submit plans and gain approval.

"On day ninety-one, I would expect each and every offending oil company to be started on actual physical cleanup, and each oil company would be given a time frame within which the job must be completed.

"Failure by any oil company to do what was required at any point along the way would result in the instant arrest of that oil company's chief executive officer on contempt of court charges. Failure would also result in the state of Alaska taking over the job and charging the oil company for cleanup costs plus heavy fines.

"The reason for the court order and the arrest on contempt of court charges is very simple. Alaska, for years, has failed to act against a single instance of illegal dumping, disposal and pollution. Instead, Alaska has chosen to wait so long that the statute of limitations has expired in each and every instance. This means that past acts of illegal disposal and pollution can no longer be treated as criminal acts, so that the only avenue left for the state to pursue is a civil suit for a "spill." These acts are not spills. They are criminal acts of illegal disposal. By invoking a court order and a contempt of court procedure, it's a whole new ball game, and the state of Alaska can make a fresh new start solving an extremely serious problem.

"The lion's share of the mess on the Kenai Peninsula would be quickly cleaned up if Alaska's governor were to shut off the natural gas pipeline on the Kenai Peninsula, running from Unocal's Kenai Gas Field on Kalifornsky Beach Road. Of course, that would short-circuit Union Oil's revenue, and it would shut down Unocal Chemicals, putting three hundred and fifty people out of work. It would shut off the supply of urea to Japan, and it would leave 250,000 people in Anchorage with no heat, as well as causing a few problems for industry.

"To clean up the rest of the state, the mess on the North Slope and the Exxon mess along Alaska's coastline, I'd agree with Governor Steve Cowper's idea. Shut off the pipeline. Then we might have to reactivate Colonel Muktuk Marston's old World War II Eskimo Army to see that no one from Washington, D.C. came up to try and turn the tap back on again.

"Perhaps shutting off either pipeline and putting a few oil company C.E.O.'s behind bars may seem like a radical move, but it's time that Alaska made a radical move in order to draw the attention of the entire world to a deadly problem that has long been hidden from view behind oil company locked gates and behind oil company propaganda.

"Remember that this is the very same oil industry that spread propaganda to the entire United States during the 1970's about a non-existent energy crisis, when in reality there was an oil glut. Many Americans waited in long lines at the gas pumps sincerely believing there was a shortage when there was not. The crisis allowed the oil industry to raise oil prices and to get the Alaska Pipeline Project moving. Now, with the crisis in the Persian Gulf, they will have that opportunity again. The oil industry has lied to you. I have told it to you like it really is. In Alaska, the oil companies have not even considered the present generation, let alone future generations.

"No wonder it was necessary to put me out of business. It had nothing to do with three-to-one slopes or with six-inch lifts. The oil industry didn't want to pay the price of legal disposal. The oil industry didn't want anyone going into the disposal business on the North Slope. Neither the oil industry not the state of Alaska nor the Kenai Peninsula Borough wanted anyone to be in possession of the information I was in possession of unless that person's silence could be assured. Certain individuals had their lives made more comfortable by the industry, and when you receive a favor, you must be prepared to return that favor. I told it exactly like it was, and for that reason, it became necessary to squash me, to silence my voice."

He paused for a moment, looking around to be sure he was alone. Then, out loud, he began to think out his arguments again.

"I took that risk with both eyes open and I paid one helluva price, but at least I woke up a lot of people in Alaska as to what was really going on and the furor has never died down.

"Well, now I'm back. Betty and I have saved some money and we're not done fighting yet.

"The first thing I'm gonna do," he grinned mischievously, "is file one huge lawsuit against the state and the oil industry."

Dave could just see the papers being drawn:

"*Alaska Environmental Industries/Mar Enterprises, Inc. and Dave and Betty Brown vs. The State of Alaska, the Kenai Peninsula Borough, and the Oil Industry.*"

"Yes," he smiled to himself. "I can just see it. There'll be those who'll say I can't win. They'll snicker like they used to about Dave Brown being a dreamer—a fool—everything that they've said before.

"But I don't care, have never cared.

"All of us are temporary caretakers of the land, and it is up to us to pass it on to future generations in good condition. This should not have to be a legal obligation. It is a moral obligation. The Lord gave it to us clean. We ought to keep it that way."

Dave lifted his eyes for a moment and saw a beluga whale breach in Cook Inlet.

"Probably chasing some salmon," he mused and felt his fishing arm twitch in anticipation just as it had when he'd first come to Alaska twenty-four years before.

"Maybe—just maybe—it isn't too late."

APPENDIX A

Chemicals and other ingredients found in various drilling muds (according to the Environmental Protection Agency and oil industries' records):

chromium	calcium chloride
barium	magnesium chloride
cadmium	carbon dioxide
copper	formic-acetic acids
manganese	sulfuric acid
nickel	chloroacetic acid
lead	hydrofluoric/hydrochloric acid
zinc	gelled acids
iron	chemical retarded acids
antimony	corrosion inhibitors
mercury	arsenic compounds
aromatic fractions	surfactants
aliphatic fractions	organic polymers
sodium fluoride	gelatinous materials
silica fluoride	calcium carbomate
polyvinyl alcohol	polyacrylamide
diphyl oxide sulfinate	ethgxylate
bacteriacides	1,1,2, trichloroethylene
—paraformaldehyde	phenols
—triophine chloride	diverting agents
—chlorine dioxide	ethylene diamine
tetracetic acid	alcohols

Swanson River oilfield water analysis results, tests by industry:

Organic Acid—C16
1,1 trichloroethane
acetone
methyl chloride
2-napthalene
4-methylphenol
Alpha BHC
Aroclor (PCB's) 1016, 1242, 1248, 1254, 1260 (found in voles in the refuge, resulting in enlarged livers)
arsenic
benzene
Beta BHC
calcium

cadmium
chloride
ethylbenzene
extractable hydrocarbons C10–C20
fuel hydrocarbons (diesel)
potassium
meta-xylene
magnesium
sodium
nickel
oil and grease
ortho and para xylene
lead
phenolics
toluene
xylene
zinc

APPENDIX B

SCHEDULE I
STERLING CAPITAL IMPROVEMENTS
PURCHASE AGREEMENT
(Dated 8/31/79)

1973	Clearing and Access Special Waste Site share = .75 × $6,147.00 =	$ 4,610.00
1974	Trench Construction	6,000.00
Surveys of Land	Special Waste Site share = .75 × $770.00 = ...	578.00
1975	Improvements (Access Road) Special Waste Site share = .50 × $2,169.00 = ...	2,169.00
1975	Improvements	462.00
	Soils Investigations	6,628.00
	Fence Installation	18,874.00
1976	Improvements	1,251.00
TOTAL PURCHASE PRICE OF STERLING SITE IMPROVEMENT		$39,487.00

SCHEDULE II
STERLING CAPITAL IMPROVEMENTS
PURCHASE AGREEMENT
(Dated 8/31/79)

List of assigned assets owned by Mar Enterprises:

1)	1 ea. 3,000 Barrel Steel Tank, estimated value	$ 30,000.00
2)	1 ea. 500 Barrel Steel Tank, estimated value	6,000.00
3)	1 ea. 10,000 Gal. Steel Tank, estimated value	3,000.00
	TOTAL ASSIGNABLE VALUE	$39,000.00

APPENDIX B

SCHEDULE III
AFFIDAVIT OF
, PRESIDENT
MAR ENTERPRISES
INC.

I do hereby swear and affirm that the value of assets listed on Schedule II of the "AGREEMENT FOR PURCHASE OF CAPITAL IMPROVEMENTS AT STERLING SPECIAL WASTE SITE" is a true and accurate value and further that I have based the determination of value on results of information provided to me by industrial users and manufacturers of similar equipment in the Kenai and Anchorage areas.

I do further swear and affirm that if, pursuant to the provisions of the "LEASE AGREEMENT AND CONTRACT FOR OPERATION OF STERLING SPECIAL WASTE SITE" and the "AGREEMENT FOR PURCHASE OF CAPITAL IMPROVEMENTS AT STERLING SPECIAL WASTE SITE", the above mentioned assets are seized by the Borough to recover capital losses incurred by the Borough as defined in the two above mentioned agreements, then MAR Enterprises will pay to the Borough any monies still owed the borough under the above mentioned purchase agreement after sale of the assets listed on Shedule II of the "AGREEMENT FOR PURCHASE OF CAPITAL IMPROVEMENTS AT STERLING SPECIAL WASTE SITE".

8-31-79
, President Date
MAR Enterprises, Inc.

I attest that is personally known to me and that I have witnessed his signature on this document.

Ruth Jacobs 8-31-79
Notary for the State of Alaska Date

My Commission expires on 1-24, 19 82.

APPENDIX B

QUOTE TO ERECT
3,000 bbl TANK

Labor & Tools $10,198.90.

Tank Scaffolding & Repair &
Build New Center Pole Extra

Gasket set 2,092.72
Fob. Tulsa

3,000 BBL Tank New $16,175
Fob. Tulsa
23,800 lbs.

3,000 BBL. Lines-1978
Price
$8,526.00

FR Melott
President Oil Industries

APPENDIX C

STATE OF ALASKA

STEVE COWPER,
GOVERNOR

ALASKA OIL AND GAS CONSERVATION COMMISSION

3001 PORCUPINE DRIVE
ANCHORAGE, ALASKA 99501-3192
PHONE: (907) 279-1433
TELECOPY NO:
(907) 276-7542

January 19, 1989

Mr. Dave Brown
5801 Fiji Street
Anchorage, AK 99507

Dear Dave:

In response to your query as to what you had to do to obtain a Permit to Drill an oil or gas well, I have enclosed a copy of our regulations.

20 AAC 25.005 details the information required when submitting an application to drill. These data are reviewed by commission staff. When satisfactory, the commission issues a Permit to Drill.

Copies of the Permit to Drill are sent to other agencies. The Permit to Drill does not exempt the applicant from obtaining other permits that may be necessary before conducting the drilling operation.

For drilling permits in the Cook Inlet basin, Permit copies are furnished to the Alaska Fish and Game Department; the Department of Environmental Conservation and U.S. Fish and Wildlife when inside the moose range. Of course, there are probably other agencies such as the Corp of Engineers, the Kenai or Mat-Su Borough who need to be notified. Knowing what permits are needed before conducting drilling operations is not a responsibility of the commission.

Hope this helps you, and best of everything.

Sincerely,

C. V. Chatterton
Chairman

dlf:l/CVC/90

Enclosure

APPENDIX D

STATE OF ALASKA

BILL SHEFFIELD,
GOVERNOR

DEPT. OF ENVIRONMENTAL CONSERVATION

Telephone: (907)
Address:

SOUTHCENTRAL REGIONAL OFFICE
437 "E" STREET, SUITE 200
ANCHORAGE, ALASKA 99501

274-2533

April 17, 1985

Mr. Dave Brown
dba Mar Enterprises
Box 1660
Soldotna, Alaska 99669

RE: Permit No. 8220-BA005 Permit Modification

Dear Mr. Brown:

Alaska Statute 46.03.120(b) provides that "the Department may modify a permit if the Department finds that a material change in the quality or classification of the waters of the State has occurred."

The Department of Environmental Conservation is hereby providing this written notification that we intend to modify your permit. This modification is deemed necessary for the following reasons:

1) Monitoring of pit contents has shown the presence of hydrocarbons and heavy metals.
2) Measurements of monitoring wells has shown the existence of low level hydrocarbons and some trace heavy metals.
3) Wastewaters disposed of at the site in pits, percolate into the ground thereby allowing for the infiltration of pollutants into the groundwater.
4) Very limited data is collected by the operator which accurately identifies the waste disposed of at the site thus making it difficult to assure groundwater protection.

APPENDIX D

The decision to modify your permit is determined necessary because activities as presently practiced at your site either directly causes or contributes to the migration of waste products from the pits to the groundwater, thereby resulting in a material change to the groundwater quality. It is our intention to make modifications which will eliminate any waste liquids from leaching from the site as well as providing better inventory control of waste products disposed of at the site.

In light of the situation which exists at the Sterling site, it is our recommendation that no new liquid waste be disposed of on the premise which could in any way percolate into the ground, until such time that new modified permit conditions are established. We encourage your active participation with us in the coming six to eight weeks while we develop acceptable operating procedures and incorporate them into a permit modification.

Sincerely,

Bill H. Lamoreaux
Regional Environmental Supervisor

BHL: lmc

cc: Keith Kelton, ADEC, Juneau
Michelle Brown, Asst A.G., Anchorage
Bob Cannone, ADEC, Kenai

APPENDIX D

LAW OFFICES OF
WILLIAM BIXBY
POST OFFICE BOX 1229
VALDEZ, ALASKA 99686

TELEPHONE (907) 835-4775

May 2, 1985

Bill Lamoreaux
Regional Environmental
Supervisor
Kenai District Office
Post Office Box 1207
Soldotna, Alaska 99669

Dear Mr. Lamoreaux:

This letter will serve as the report you requested in regards to the Notice of Violation you gave Dave Brown on April 17, 1985. The following is our response to each alleged violation:

Item II, D, 1: Pond and dike construction not of three to one slope. According to an agreement between your agency and my client, this matter did not have to be completed until June 20, 1985. See also James W. Cross' letter of April 22, 1985 and his letter of April 6, 1984.

The alleged violation that diversion constructions were not constructed around ponds, is without merit. These structures were completed on April 5, 1984, and at the time of the March 5, 1985 inspection by yourself and Mr. Kross, Mr. Kross noted that this alleged violation had been rectified. See Mr. Kross' letter of April 22, 1985.

The alleged violation that the as-built plans for the diversion structures were not submitted to you is also without merit. The as-built plans were submitted to Bob Martin of your agency on September 5, 1984.

In regards to the alleged violation of Item III, D,

5, that drilling muds were not mixed with soils, compacted or covered, there are two responses. The first is found in Jim Kross' letter of April 22, 1985. Mr. Kross has informed me that the site plan of operations submitted in December of 1983 specifies a new technique for the disposal of drilling muds at this site. Please see Mr. Kross' letter of that same date and his attachments. Further, I specifically recall in August of 1984 sitting in Governor Sheffield's conference room. Also present were Bob Martin, of your agency, Keith Kelton, the head of the Solid Waste Unit, Ray Gillespie, of Governor Sheffield's office, Lenny Boston, of Governor Sheffield's office, Tom Janke, of the Attorney General's office, John Parker, my investigator, and Howard Kaiser, an ex-employee with the Alaska Department of Environmental Conservation. I recall specifically addressing this issue to Bob Martin and Keith Kelton. I questioned Mr. Martin and Mr. Kelton regarding Union Oil's permit application for their seven proposed disposal sites. I stated that there were no provisions for Union Oil to mix their drilling mud with soil and compact them in lifts as my client is required to do both by regulation and his permit. I was told by Mr. Martin that Mr. Brown was no longer required to do this. Further, Mr. Kelton stated that the regulation was only applicable to sanitary land fills and not applicable to Mr. Brown's operation.

In regards to the alleged violation including Item II, A, failure to follow a plan of operation as specified in Solid Waste Management permit application amended October 28, 1981. I have been informed by Jim Kross that the new plan of operation previously submitted to your office supercedes by agreement with your agency, the plan of operation of October 28, 1981. Further, in my letter to you dated April 26, 1985, you agreed that this alleged violation was too indefinite for me to respond to. You agreed that you would check with your field officers and state the specifics as to how the plan of operation was not being followed. Further, you agreed to an extension of time that would allow us reasonable time to adequately respond to your new and specific allegations.

Having reviewed the correspondence and having discussed this matter with Mr. Kross, it is apparent that there is a lack of communication between members of your agency. I suggest that this matter can be rectified by reducing agreements and conversations to writing. Mr. Kross has told me that he has attempted to do this. He states that Mr. Martin of your agency will never put their agreements in writing nor does he respond to Mr. Kross' letters. If this were the case, misunderstandings such as your April 17th Notice of Violation should be avoided in the future.

If you have any questions or desire any further information about these alleged violations do not hesitate to contact me. We would like to get this matter straightened out with your agency.

Best regards,

LAW OFFICES OF WILLIAM BIXBY

William Bixby

WB/ldb

attachments.

APPENDIX E

STATE OF ALASKA

BILL SHEFFIELD,
GOVERNOR

DEPT. OF ENVIRONMENTAL CONSERVATION

Telephone: *(907)* 262-5210
Address: P O Box 1207
Soldotna, AK
99669

July 11, 1985

Mr. Dave Brown
Alaska Environmental Industries
Box 1660
Soldotna, AK 99669

Re: Sterling Special Waste Site

Dear Mr. Brown:

This is to request a copy of the waste disposal log for the Sterling Special Waste Site from January 1, 1985 to the present.

Also, this office recently received a complaint regarding steam or gasses being released from your shop facility in North Kenai from a nearby resident by the name of Ms. Rita Johnson. Could you please explain to us the nature of these steam or gas releases, so that we can respond to Ms. Johnson's complaint.

Thank you for your assistance.

Sincerely,

Environmental Field Officer

at

APPENDIX E

ALASKA ENVIRONMENTAL INDUSTRIES, INC.

Industrial Vacuum Pumping
Waste Disposal
Tank Cleaning
Tank Liners
Oil & Chemical Spill Cleanup

Formerly Mar Enterprises
Box 1660 • Soldotna, Alaska 99669

Bus: 262-9073 Or 262-4875
Res: 262-4798

Power Washing
Roustabouts
Power Washer &
Steam Cleaner Sales,
Service & Rentals

July 30, 1985

ADEC
Soldotna, Alaska

In answer to your recent inquiry regarding smoke or steam on our property, It was steam.

In the normal course of our business while cleaning tanks whether they be vacuum tanks, Sealand tankers, Oil company tanks or whatever, we use hot high pressure washers, and we also use steam.

We have done this for the past five years, with the D.E.C. taking no interest ever before.

Its strange that the lady would call the D.E.C. rather than the Fire Department or the Police Department.

I hope this answers your questions.

Sincerely,

Dave Brown

APPENDIX F

STATE OF ALASKA

BILL SHEFFIELD,
GOVERNOR

**DEPT. OF ENVIRONMENTAL CONSERVATION
ENVIRONMENTAL QUALITY OPERATIONS**

Telephone: (907) 274-2533
Address: 437 E Street
Suite 200
Anchorage, AK
99501

CERTIFIED MAIL
RETURN RECEIPT
REQUESTED

July 6, 1983

Mr. Dave Brown
MAR Enterprises
P.O. Box 1660
Soldotna, AK 99669

Dear Mr. Brown:

RE: Permit Violation (Sterling Special Waste Site)

The Department of Environmental Conservation (ADEC) has yet to receive the July 1, 1983 quarterly monitoring results as required in IV B.3. of your permit No. 8220-BA005. This is especially alarming in that we had previously addressed our concerns on similar episodes in our letters of December 27, 1982 and May 25, 1983.

The importance of this routine sampling in the detection of any possible contamination of the underlying aquifer cannot be over emphasized. Your failure to provide sample results on the required schedule has hampered our ability to evaluate whether there are any adverse impacts of continued operation. You are thus in non-compliance with specific conditions of your waste disposal permit.

Therefore, until such time that required samples are submitted and a written response received from this office, you are to immediately discontinue receiving any additional waste materials at the MAR site. Further, this does not relieve you of other actions by this Department for permit non-compliance which may be instituted by this Department.

Sincerely,

Bob Martin, P.E.
Deputy Director

BM/jfr/gk

cc: Commissioner Richard Neve
Bob Flint
Ron Rice
Keith Kelton
Bob Cannone
Mayor Stan Thompson
John Wood

APPENDIX G

ALASKA ENVIRONMENTAL INDUSTRIES, INC.

Industrial Vacuum Pumping	Formerly Mar Enterprises	Power Washing
Waste Disposal	Box 1660 • Soldotna, Alaska 99669	Roustabouts
Tank Cleaning		Power Washer &
Tank Liners	Bus: 262-9073 Or 262-4875	Steam Cleaner Sales,
Oil & Chemical Spill Cleanup	Res: 262-4798	Service & Rentals

STERLING SPECIAL WASTE SITE
MONIES INVESTED

Well Drilling- (Spires)	$12,993.00
Dirt Work-	
(Jess Robinson)	22,970.65
(Jackson)	8,146.98
Gravel	226.85
Engineering & Consulting	
Dowling & Rice	$15,671.00
Environmental Services	32,807.12
McClelland E.B.A.	1,000.00

Dirt Work by A.E.I. using
AGC Equipment Rental Rates

3 YD Hoe W/ Operator-	$150.00 hr.
D3 Cat Dozer Wide Pads	60.00 hr.
	$210.00 hr.

Winter of 82-83	126 hrs.	$26,460.00
Winter of 83-84	120 hrs.	25,200.00
Summer of 84	160 hrs.	33,600.00
Winter of 84-85	41 hrs.	8,610.00
Backhoe Welding-Jetov		5,570.00
	Rick Alton	345.00
Backhoe Swing Motors		6,000.00
Moving Backhoe (Browns Const)		1,345.65
Moving Grader (Hustlers)		900.00
Supplies dump site 1985		1,063.81

APPENDIX G

Parts & Labor on Equipment			
1983		7,027.00	
1984		5,976.00	
Dewatering 1985			
Bud Williams		3,576.00	
John Pike			
Carrie Wright		893.00	
Rick Colby		4,800.00	4 mon. @ 1,200.00mon
Wages 80-83		69,500.00	Misc. before computer & no breakdown
Sub-Total		294,682.06	
Monitoring Costs	80&81	$5,682.00	
	82	6,000.00	
	83	6,501.25	
	84	5,045.75	
	85	6,915.27	
	86	3,000.00	
Borough Taxes		5,500.00	
Borough Lease		26,240.00	
Insurance	1983	7,289.00	
	1984	8,571.25	
	1985	10,083.00	
	1980, 81 & 82 EST.	20,000.00	Ins. lumped w/ other ins. those years.
		$110,827.52	
	From page 1	$294,682.06	
	Total	405,509.58	

APPENDIX H

ALASKA ENVIRONMENTAL INDUSTRIES, INC.

Industrial Vacuum Pumping
Waste Disposal
Tank Cleaning
Tank Liners
Oil & Chemical Spill Cleanup

Formerly Mar Enterprises
Box 1660 • Soldotna, Alaska 99669

Bus: 262-9073 Or 262-4875
Res: 262-4798

Power Washing
Roustabouts
Power Washer &
Steam Cleaner Sales,
Service & Rentals

ALASKA Department of
Environmental Conservation
437 E Street
Second Floor
Anchorage, AK 99501

Re: Union Oil Company
Disposal Permits
8423-BA008 and
8423-BA009

Gentlemen:

I wish to object to the issuance of any Solid Waste Permits to Union Oil Company, and particularly to permit numbers 8423-BA008 and 8423-BA009, for which they are currently applying.

My objections are long overdue, valid, and are based upon true facts gained from my own observations, not hearsay.

The State of Alaska has very clear cut regulations pertaining to the disposal of solid waste, and drilling mud is solid waste.

The Kenai Peninsula Borough and the city of Kenai have a few solid waste regulations of their own.

Time and space don't allow me to go into an in depth discussion of these various regulations. I will discuss only one.

The first and most important State Solid Waste Regulation is the one which states that Solid Waste can not be buried or disposed of without a permit. The waste must go to an approved waste site, not 3 years later, but immediately.

The only site on the entire Kenai Peninsula that is approved for the disposal of drilling mud is the Sterling Special Waste Site.

My wife and I operate this site under a lease from the Kenai Peninsula Borough, which does not expire until 9/1/98, and under permit number 8220-BA005, issued by Alaska Department of Environmental Conservation.

APPENDIX H

My first experience with Union Oil Company was in May 1981, when at Jim Callendar's request, I personally measured the amount of drilling mud located at their Cannery Loop 2 reserve pit. I wrote Mr. Callendar on May 22, 1981 and advised him that his pit contained approximately 50,000 barrels of drilling mud, and I gave him my price for removing all the mud from the reserve pit, and transporting it to the Sterling Special Waste Site for disposal. He then contracted with me based on the information I supplied to him to remove all the mud from the Cannery Loop 2 reserve pit. This contract was issued using an hourly rate instead of a flat rate.

The dimensions of the Cannery Loop 2 reserve pit when I measured it in May 1981 were 400 feet long by 80 feet wide, and it contained approximately 8 foot of mud therein. A few days later Mr. Callendar contacted me by phone and advised me to start removing the mud from the pit. When I arrived the pit that two days earlier was 400 feet long had suddenly shrunk to a mere 200 feet. Union Oil Company had backfilled approximately 50% of the pit before we started pumping, and they continued backfilling as we pumped what was left.

On 6/10/81 I again wrote Mr. Callendar and advised him of the total amount of mud removed from Cannery Loop 2 as well as the costs to remove it. We disposed of 15,823 barrels of drilling mud from Union Oil's Cannery Loop 2 reserve pit. The other approximately 34,000 barrels were buried by Union Oil Company, on location, and against at least DEC regulations, with no permit. This is in a residential area on Beaver Loop Road in Kenai, which has a high water table, and shallow drinking water wells.

After Cannery Loop 2, Union Oil Co. moved closer to Kenai and drilled Cannery Loop 3. This well is located on the bridge access road across from the Chevron Oil Bulk Plant. Out of this well we disposed of approximately 8,000 barrels of used drilling mud, the rest to remain behind a chain link fence to slowly leach into the local water table.

After Cannery Loop 3 Union Oil Co. moved to the Kalifonski Beach Road Gas Fields. In the Gas Fields I am going to estimate that they have drilled approximately 10-12 more wells. Union Oil Co. has received 15 permits for on shore drilling for 1980-84, and I am going to assume that two or three of these have yet to be drilled. They have also received a permit for an injection well, but evidently it does not work well as they are still using mud pits everyday. Assuming a total of twelve wells drilled which each used approx. 40,000 to 50,000 barrels of drilling mud we are looking at approximately 500,000 barrels of drilling mud disposed of, mostly in a swamp where it's in the water table as soon as the vacuum truck blows it's load off.

Anyway out of all these additional wells drilled, we received a total of slightly over 11,000 barrels of drilling mud, and if you add that to 23,823 barrels of drilling mud removed from the two Cannery Loop wells you come up with 34,823 barrels of drilling mud, which is less than the average amount that should be produced in one well.

Union Oil Company has in the past three and half years disposed of, with no permit, approximately 500,000 barrels of used drilling muds. This is 21,000,000 gallons. Where is it all going? Certainly not to the Sterling Special Waste Site where it is supposed to be going. They are now asking the Alaska Department of Environmental Conservation for an after the fact permit to leave there what was illegal in the first place. The area they are requesting a permit to dispose in has a high water table and poor soil conditions for this type of disposal. No permit could be issued by any individual or agency without making a 180 degree change in the existing regulations.

Union Oil Company was advised by Howard Keiser (who was with the DEC at that time), by letter dated 5/4/79 and 12/7/81 that they were in violation of DEC regulations if they did not dispose of their drilling mud at the Sterling Special Waste Site. He started an investigation, but retired in 1982. To date there is no indication that DEC ever followed up on that investigation.

In early July 1984 Howard Keiser and I walked through the Union Oil Gas Field marshes with hip boots, cameras, and binoculars. These marshes are among the most beautiful in the state, and they have always been a nesting and resting ground for many species of water fowl. Last year the bird numbers had decreased noticeably. This year there doesn't appear to be any birds in the area at all. We saw one duck, a lone pin-tail, and he was only passing through. There were no geese at all. There were no terns. There were no sea-gulls. There were four sandhill cranes, and four or five sandpipers. The cranes had no eggs and no babies, because they made none of the threatening gestures that cranes make under these circumstances. They are smarter than we are, so when Union Oil Company fouled up their environment they simply left.

There is an intermittent oily sheen starting at Union Oil's last mud pit and extending out about one mile into this beautiful marsh. There is drilling mud up three and four feet high on tree trunks near Union Oil's last mud pit, indicating that the mud pit has been allowed to overflow.

It appears that drilling mud is not the only thing being covered up. We as tax payers are paying the DEC to protect our environment, while for some reason, which they don't care to discuss, they are deliberatly looking the other way, and allowing our environment to become destroyed.

If our police didn't protect us from crime, if our firemen didn't protect us from fires, and if our teachers didn't educate our children, then Alaska would be a terrible place to live. Is our environment any less important?

Bank robbery would become a much more popular sport than it now is, and Robert Hansen would still be running loose, raping and murdering women, if only the police would look the other way.

I don't think that Union Oil Company has any right to bend the rules to suit their purposes. I don't think that Union Oil Company has any right to ask for an after the fact permit to violate the law, and make legitimate that which never was legitimate.

This is merely the tip of the iceberg. During the past four years, not merely 500,000 barrels of drilling mud has been buried illegally with no permits. The total in four years has been in excess of 1,000,000 barrels covered up where it shouldn't be. Union Oil Co. is not the only offender. They are merely the worst offender.

APPENDIX H

There are at the present time in excess of one hundred illegal dump sites on the Kenai Peninsula, untouched oil spills, waste oil dumpings, septic wastes on top of the ground, land fills that are a disgrace, and not operating even close to the parameters of their permits.

Don't say DEC, that you didn't know. You did know. You didn't care!

I expect that as a result of my revelations, the DEC will retaliate, and attempt to shut the Sterling Special Waste Site down, because I didn't dot my i's or cross my t's. Union Oil may have already retaliated, because they have already cancelled a platform job we had with them.

I have already told the Alaska State Ombudsman, Jack Chenoweth, to expect this. I am putting this entire matter in his hands. If he can't handle it, then if necessary I'll go to the US Attorney General. I have faith though in Jack.

I expect that my revelations are going to make me very unpopular in a few circles. That's OK. I came here to run a business, not to win any popularity contest.

As Harry Truman said, "If you can't stand the heat, then stay out of the kitchen."

I think we Alaskans as citizens, residents, taxpayers, and voters, have a moral obligation to demand the removal of an official, elected or appointed, or of and civil servant who refuses to do the job which he is paid to do.

Let's have a public hearing on this. Let's have no retroactive permits issued. Let's demand to know why the DEC is doing nothing.

Any one wishing more information, please feel free to call me, but your objections must be delivered to the Anchorage DEC office by Oct. 29, 1984, or you may object personally at the public hearing in Kenai on Oct. 29. Phone 283-5050 or 283-4906.

Sincerely,
Dave Brown
Alaska Environmental
Industries, Inc.
Keeping Alaska Clean

APPENDIX H

P.S.

To the people of the Kenai Peninsula.

Chances are you moved down here for an improved life style, to get away from the dirt and the filth in the cities.

In case you haven't already noticed, we are starting to get somewhat dirty and filthy down here too.

The ducks, the geese, the terns and the sea gulls are already moving on to other areas.

If you think our local moose all die of old age, highway accidents, or hunting accidents, then you are wrong. Some of them die by falling into drilling mud pits.

If you live in one of the following affected areas, you may eventually notice a difference in the taste of your drinking water, if you haven't already:

Beaver Loop Road
Kenai River Bridge Crossing Road
Kalifornsky Beach Road
Cannery Road

Come to the meeting on October 29 in Kenai, stand up and be counted!

APPENDIX I

ALASKA ENVIRONMENTAL INDUSTRIES, INC.

Industrial Vacuum Pumping
Waste Disposal
Tank Cleaning
Tank Liners
Oil & Chemical Spill Cleanup

Formerly Mar Enterprises
Box 1660 • Soldotna, Alaska 99669

Bus: 283-5050 or 283-4906
Res: 262-4798

Power Washing
Roustabouts
Power Washer &
Steam Cleaner Sales,
Service & Rentals

The People of the Kenai Peninsula:

Re: Union Oil Company
Disposal Permits
8423-BA08 and 8423-BA009

Residents:

I have evidence to indicate that on the Kenai Peninsula some extremely serious dirty tricks may have been played between certain Oil Companies and certain employees (not all of them) of the Alaska Department of Environmental Conservation.

I said may have been. They also may not have been.

In the United States of America, the accused is considered innocent until proven guilty.

At this point no one has been accused. There is enough evidence to warrant an investigation.

For the last four years as the operator of the Sterling Special Waste Site, I have been deliberately and ridiculously harrassed by the D.E.C. and I have been subjected by them to what I call, "Selective Enforcement of the Law," I.E.: You do this Dave Brown, but no one else has to.

This in itself is a violation of my Constitutional Rights under the 14th ammendment to the U.S. Constitution, which says no state shall deny to any person within its jurisdiction the equal protection of the laws, "Equal Protection Under the Law".

The reasons for which I am being harrassed and selectively enforced are very simple.

1.) I don't treat the D.E.C. with the respect which they feel they are entitled to.

My reason: So far I have met no one employed by the D.E.C. whom I felt was deserving of my respect.

2.) I know too much. As the operator of the Sterling Special Waste Site, with the records I keep, I know who is obeying the rules, and I know who is violating them. When I ask the D.E.C. why these violations are allowed they refuse to answer me.

3.) I am very outspoken. I tell it exactly like it is and I let the chips fall where they will.

My evidence indicates that certain Oil Interests may have made it worth while for certain D.E.C. people to "close their eyes" , and let the law be violated, in order to save hundreds of thousands, even millions of dollars for these oil interests.

If this is the case, then the parties involved should be tried by the Court System.

If this is the case then we do not have "government of the people, by the people, and for the people. Instead we have "government of the people, by the oil companies, and for the oil companies."

If the D.E.C. could put me out of business, then I would no longer have all the facts, and my voice would no longer be effective. I could not tell you, the people of the State of Alaska, what I am telling you now.

I can not speak for all of the oil companies. My dealings primarily are with those oil companies who have done onshore drilling during the past four years.

Regarding on shore drilling and the companies connected with drilling, let me say this:

Any oil field service company, excavating contractor, or vacuum pumping service, that knowingly assists an oil company in performing AN ILLEGAL ACT, is just as guilty as the oil company itself. The service company is an accessory to the illegal act.

We have laws, statutes, rules and regulations. These were not laid down or enacted, to save money for the oil companies, nor to be broken by the oil companies, with or without the help of civil servants, lobbyists, or politicians.

These laws and regulations were laid down for the protection of the Environment of the State of Alaska, for the people of the State of Alaska, and in the case of the Kenai National Moose Range, for the people of the United States of America.

We need someone close by to watch the D.E.C., because as has already been proven they are inept, incapable, and unconcerned, for whatever reason.

We have federal regulations, state regulations,

borough regulations, and in some cases municipal regulations. No single agency has the right to allow any other agencies regulations to be violated.

Since the Kenai Peninsula Borough boundaries are far reaching, and since they have offices in Soldotna with an Assembly and a Mayor that are always accessible, then I say lets let the Borough get involved. Lets not be restricted to dealing with D.E.C. people who are always either uncooperative or inaccessible.

Recently the Borough has decided to establish a Solid Waste Commission, a long overdue good idea. Lets consider having that commission involved in Solid Waste & Liquid Waste. Lets give them the power to investigate nasty situations. Then lets give them a free rein to go to the D.E.C. with the problem, and then to go over the D.E.C.s head to the proper higher authority, if the D.E.C. refuses to act immediately.

I am very favorably impressed with what I see on the Kenai Peninsula Borough Assembly, people helping people for the good of all of us.

I am not impressed favorably at all with the D.E.C. and never have been. We need a housecleaning in that department.

Even the "Golden Rule" has been polluted to where it now reads, "He who has the gold makes the rules" .

If we the people don't make some changes in the system as it now exists, then twenty years from now if you want to show your kids a moose, you'll have to take them to the Childrens Zoo in Anchorage.

If you want to keep our ducks, our geese, our terns, our sea gulls and our moose, if you want to keep our water clean, if you want to restore our Golden Rule to what it was and what it should be, then come to the Public Hearing on October 29, 1984 regarding Union Oils Disposal Permit Request. Time and place: 7 p.m., Kenai City Hall, 210 Fidalgo St., Kenai.

Call me if you have any questions, work: 283-5050 or 283-4906, home: 262-4798.

Thank you and I'll see you there.

Sincerely,
Dave Brown
Alaska Environmental Industries, Inc.
Keeping Alaska Clean For Your Kids and Mine

P.S.: To the D.E.C. If you don't like what I have to say, thats understandable. However, I would be happy to take a polygraph test on any and all subjects concerning AEI, the Sterling Special Waste Site, and or Dave Brown personally, back as far as 8/24/26, my date of birth. I would be exceedingly happy if certain D.E.C. employees would be willing to do the same. I would be happy to personally pay for all of the tests, D.E.C.s and mine.

APPENDIX J

LAW OFFICES OF
WILLIAM BIXBY
POST OFFICE BOX 1229
VALDEZ, ALASKA 99600

TELEPHONE (907) 835-4775

Oct. 26, 1984

Dear Kenai Peninsula Residents:

Attached to this letter is a copy of an article from the October 23, 1984 issue of the WALL STREET JOURNAL regarding the disposal of drilling mud in Louisana. This timely article describes a true horror story resulting from the improper disposal of drilling mud. Unfortunately you & your family may face the same danger as the people described in this article. THERE ARE AT LEAST 100 KNOWN & ILLEGAL DISPOSAL PITS IN THE KENAI AREA.

Big oil money and inefficient government have teamed up here in Alaska as they have in Louisana. The result could possibly kill you or your loved ones. Union Oil Co. has applied for a permit to dispose of this potentially lethal waste in your "backyard" .

On Monday night Oct. 29, 1984 at 7:00pm, a public hearing will be held at Kenai City Hall, 210 Fidalgo st., Kenai, on Union Oil permit application. If you do no want Union Oil to engage in this activity please attend this hearing and strongly voice your opposition to the granting of this permit. If you can not attend please call Governor Sheffield at 465-3500 to voice your opposition. I would also suggest that you contact your local Borrough Assembly Representative, State Legislators and Senators.

Thank you for your time and consideration of this important issue.

Best Regards,

William Bixby

APPENDIX J

THE WALL STREET JOURNAL

TUESDAY, OCTOBER 23, 1984

Oil's Legacy

In Louisiana, Pollution And Cancer Are Rife
In the Petroleum Area

Region Is Littered With Pits
Of the Chemicals Poured
In Wells During Drilling

A Small-Town Doctor's Tale

By THOMAS PETZNGER JR.
and GEORGE GETSCHOW
Staff Reporters of THE WALL STREET JOURNAL

BATON ROUGE, La. —At the lake next to the governor's mansion here, "no fishing" signs have sprung up. The Department of Environmental Quality posted them. The governor's lake contains PCBs.

It isn't unusual to find contaminated water here in the state's industrial, oil-refining and petrochemical corridor. Tons of waste containing potentially toxic heavy metals and organic chemicals are generated daily, and have been for decades.

But Louisiana's water problems aren't confined to the industrial zones. In one town on the coastal marshes of South Louisiana, a test of 84 water wells found that 81 contained heavy metals. The water in another community is so briny that drinking fountains are stained white. In some swamps, groves of cypress stand dead.

These are areas of heavy oil and gas production, where thousands of open waste pits brim with brine from far underground and with the soup of chemicals poured into wells during drilling. Other waste sludge is sunk in old wells for disposal, spread over adjoining land or, if the drilling site is offshore, simply dumped into the water.

Some energy and petrochemical companies in Louisiana have begun to take more care in disposing of wastes. A few are spending millions to clean up their worst dumps. But much waste continues to be disposed of haphazardly, and cleanup efforts can do little if a decades-old waste pit has leached down to an aquifer below.

Meanwhile, unusual levels of illness are showing up in much of South Louisiana. No one can demonstrate whether any of the illness is linked to oil and gas activity or to the area's extensive water pollution. But concern is mounting, and some of the cases are

The Vincent Case

The area around Kaplan, La., where Jerome Vincent lives, is a site of many old and new oil and gas wells. Superior Oil Co. sank another one, a very deep dry hole, in the rice field behind his home four years ago. It left behind some leaky barrels and three open waste pits.

In May 1983, a private laboratory tested Mr. Vincent's well water and found it contained chromium, a metal that is sometimes dumped in oil wells during drilling and that toxicology manuals link with cancer. His well's chromium level was slightly higher than the federal limit for public drinking water systems.

Tests by the state health department found chromium-rich sludge in Mr. Vincent's water-well pipe. The private lab tested sludge from one of the waste pits and found its chromium level to be more than 2,700 times the concentration in Mr. Vincent's water.

In June 1983, Mr. Vincent's 40-year-old wife died of leukemia. Two months later, his 62-year-old father, who lived next door, died of pancreatic cancer. Then this year his 21-year-old son died of cardiopulmonary arrest associated with fatty infiltration of the liver.

Superior, now a Mobil Corp. unit, eventually hired contractors to clean up its pits. It says in a written response to an inquiry that its records "do not reveal any evidence of deaths or illness" from oil-field contamination in Louisiana

Another Cluster

In adjoining Cameron Parish (county), Annette Baccigalopi worries about an oil-waste disposal well a few hundred yards from her home, a well the state temporarily closed this year because of possible leaks. Her anxiety seems understandable; in the past several years, cancer has taken the lives of her father, her father-in-law, her brother-in-law and her daughter's father-in-law, all from the same area. Her mother-in-law has cancer now, and her mother and a cousin have had cancer operations.

Down the road, Debbie Theriot says that recently, within three weeks, her mother died of pancreatic cancer and her father of leukemia.

Statewide, cancer deaths are 9.1% higher than the U.S. average, using the figure for white males. The problem is more severe in the major oil producing and refining region of South Louisiana. Ten counties there rank in the top 5% nationwide for cancer deaths per capita among white males. A 1980 study published in the Journal of Occupational Medicine showed that pumpers, oil-field laborers and certain refinery workers were more than twice as likely to get lung cancer as other South Louisianians. (Smoking rates appeared to be similar.)

Yet the state hasn't made any systematic effort to determine whether South

Louisiana's health problems are linked to the concentrations of oilfield waste. And what is known isn't always passed along to residents. For instance, state tests last year found the ground water in the town of Port Barre to be of questionable quality, but residents weren't told about the tests.

It is hardly surprising that Louisiana should have a water problem, given the volume of harsh chemicals spread about during oil and gas activity. Consider what happens whenever a new well goes in.

Chemical Cornucopia

There is much more to drilling an oil or gas well than digging a hole and lowering a pipe. Throughout the process, operators inject into the hole tons of viscous material called drilling mud, a blend of clays and weighting agents that flushes cuttings to the surface and keeps the hole from collapsing. The "mud" may contain any of 500 different chemicals that act variously as foaming agents, defoamers, flocculants, thinners, viscosifiers or emulsifiers.

The chemical cornucopia often includes such materials as asbestos, formaldehyde, carbolic acid and caustic soda (the active ingredients in Drano). Among the potentially toxic metals that drilling mud may contain are barium, chromium, arsenic, aluminum, titanium, and zinc. To prevent high-pressure formations from blowing out, lead or iron may be added to increase the weight. To reduce pipe and drill-bit friction, such lubricants as asphalt laced with phenols may be added.

If the well is a producer, for years afterward the operators administer regular doses of corrosion inhibitors, such as ammonium bisulfite and zinc chromate, and anti-bacteria agents, such as acid compounds and formaldehyde.

Nowhere is the use of these chemicals more extensive than in the deep-pocket formations of Louisiana, with their high temperatures, intense pressures and unusual strata. "There's more money in this business (here) than anywhere else in the world," says Tom Parker, a manager for a unit of NL Industries Inc., a leading maker of drilling mud.

The state's oil and gas industry spends about $1 billion a year on drilling compounds, more than the total for the rest of the country. In South Louisiana, where wells tend to be especially deep, drilling a typical one of 10,000 feet might require a million pounds of "mud." In all, oil men drill about 4,000 new oil and gas wells in the state every year.

As much as one-third of the drilling mud poured in remains underground. The rest returns to the surface for recycling or disposal. What comes up may be even harsher than what went down, as it can pick up naturally occurring substances like mercury, antimony, cobalt, cadmium, thallium, selenium and radioactive isotopes of potassium.

If a well is a dry hole, any used drilling mud that can't be recycled is commonly dumped into it. But if the well is a producer, hauling away all the waste material could easily cost $100,000. As a result, over the years much of it has simply been abandoned in earthen pits of an acre or so. In addition, oil and gas companies often use smaller pits to hold waste oil and other fluids during the life of the well. An estimated 13,000 oil-field pits litter the Louisiana landscape today, an unknown number of them leaching down to underground water supplies.

Often an oil company spreads the used drilling mud over the nearby land. "A lot of times they're spreading stuff that shouldn't be spread at all," says Jerry Fontenot of Lafayette, an agronomist who has examined hundreds of sites for farmers and oil companies alike. "They move off, but the farmer is left with infertile soil for years." Industry studies show that plants can take up heavy metals from drilling mud spread on the land.

No Federal Oversight

The federal Environmental Protection Agency once proposed to regulate drilling wastes. But lobbyists for the oil industry contended such regulation would cost it $45 billion in the first year, and Congress in 1980 exempted oil and gas waste from regulation. It instead ordered the EPA to study whether such waste posed a public health hazard, but the EPA has never done the study. The agency "doesn't consider it as high a priority as other projects," such as studies of utility or mining waste, explains John Heffelfinger, an EPA official.

Louisiana doesn't classify drilling waste as hazardous, and the industry says this only makes sense. William Bailey, an official of Mid-Continent Oil & Gas Association, a trade group, says that while he wouldn't want to "pick it up and make a sandwich" of the stuff, "I'm not aware of anything in drilling mud that's hazardous."

But another major oil-producing state, California, takes a far different view. It requires that all drilling compounds with certain levels of metals and other toxic materials be disposed of as carefully as wastes classed as hazardous. David Storm, a California state toxicologist, estimates that as much as 25% of what is in drilling mud may be hazardous. Drilling compounds, he says, "pose a very complex problem in terms of the number of chemicals used, the types of chemicals and the number of drilling sites they're generated at. The potential hazard of oil-field waste has been understudied because of the complexity of the problem."

What research the industry has done isn't especially reassuring. A study sponsored by the American Petroleum Institute sampled ground water at some drilling-waste sites around the country and found arsenic, barium, chloride, zinc, cadmium, sodium, lead and mercury in the water at every site.

At a Shell Oil Co. drilling-mud waste pit in South Louisiana, the researchers found arsenic in the ground water at roughly twice the federal limit for public drinking water, as well as lead and mercury at roughly three times the limit. The "mud" used in drilling the well contained 1,600 pounds of asbestos.

Nevertheless, the study concluded the drilling-mud pits don't present a hazard to human health or the environment. Some of the ground-water contamination occurs very slowly, and the rest usually declines to safe levels within "several hundred feet" of the pit, the study said. It also speculated that some of the heavy metals might have come from pesticides or fertilizers spread on nearby farms.

Louisiana's streams and forests are also common dumping places. In 1980, a hauler of oil-field waste was caught dumping a load into Five Mile Creek near Ruston, La. When he tried to cover up by setting fire to the waste, the blaze raced down the creek and destroyed a bridge. In another case, last year a waste-hauling firm operated by Sandra Thompson, a former head of Louisiana's Tourism Board, Trails Council and Beautification Commission, was fined for dumping drilling mud into a wooded area that drains into the West Pearl River. She says she wasn't aware of the illegal dumping.

Off the coast, drillers dump much of their used drilling mud and rock cuttings into the sea, a practice they maintain is harmless to aquatic life. But an executive of Hughes Drilling Fluids, a major supplier, writes in a recently published paper that small quantities of one common drilling lubricant, diesel fuel, can cling to cuttings even after they are washed and kill the Mysid shrimp, "a basic link in the complex food chain of marine life." Hughes and other suppliers offer special fluids for "environmentally sensitive" areas, but they cost more and make up only a small part of sales.

Louisiana's aging oil fields also produce large volumes of salt water—an average of eight times as much as oil or gas. This water from deep underground is saltier than the ocean, sometimes seven times as salty. It also contains naturally occurring but toxic substances such as arsenic, as well as some of the organic chemicals found in crude oil; the latter chemicals "create the most significant toxicity problems," according to a report conducted for the Gas Research Institute, a trade group.

Some of this brine is dumped (illegally) into streams, marshes and roadside ditches. But most of it now is injected into disposal wells, of which the state has about 4,500. Louisiana didn't begin monitoring disposal wells until 1982 (19 years after Texas began doing so); when it did start, it found 11% of those checked to be unsound, unpermitted or in violation of the new rules. One inspection report said a certain household's drinking water was tainted by a nearby disposal well that was only a pipe in the ground, without any casing to protect ground water.

Cleaning Up Pits

Another brine-disposal method, the large open pit, has been regulated for 15 years in Texas but still isn't in Louisiana. Fritz Spencer, a career regulator who is holding hearings this week on a proposal to regulate the pits, says that about three-fourths of major oil companies have stopped using pits for brine except in emergencies. And some have programs under way to clean up waste pits.

Exxon Corp. says it has about 300 oil-field pits, some decades old, but when it leaves a production site it clears the pits. It says it doesn't have any abandoned pits. Standard Oil Co. (Indiana) says it has about 100 "temporarily abandoned" pits at any time but cleans them all up within a year or two. Gulf Oil Corp. says that for the past eight or 10 years it has been lining its brine pits with vinyl to prevent leaching; it says it doesn't know how many abandoned pits it has but is currently trying to find out.

Among cleanup efforts, Mobil Corp. says it has spent $450,000 to reclaim 12 old pits so far and has budgeted money to close all of its pits at production sites. Texaco Inc. said it will spend $11 million through 1985 to clean up more than 200 pits in Louisiana, all of which "could pose a potential environmental problem."

Still, Mr. Spencer, the regulator, considers the industry's efforts only a small start. "There is an extreme need of remedial action on many, many pits," he says.

Not all state officials share this level of concern. Gov. Edwin Edwards, echoing the view of many in the industry, says that "you could live a long, happy life" next to a pit of oil-field waste.

Farmers' Complaints

But W.T. Langley, a big grain and soybean farmer in South Louisiana, claims that drilling mud spread by Standard Oil Co. of California left 10 acres of his farm land useless. He has sued, asking that Socal replace the tainted dirt with new topsoil. Socal maintains in a court filing that the damage wasn't out of the ordinary.

Marlin Dupre's problem was salt water. The ex-farmer from Franklin, La., says rusty pipelines that Indiana Standard uses to link oil wells with disposal wells leaked on his sugar-cane land for years, damaging more each year. Two years ago, he says, with 250 acres poisoned, he gave up and quit farming. Indiana Standard doesn't deny that saltwater damage occurred but won't comment further.

Mr. Dupre says such contamination "is all through South Louisiana where I see oil production. What kind of price did Louisiana have to pay to keep the rest of the nation warm?"

Neighbors of oil-field waste pits often tell of being left to deal with such problems after an oil company moves out. "We complain and complain about these smelly pits, but no one will take responsibility for them," says Marion Sellers of Abbeville, La. He built a two-foot-high levee around his house in an effort to keep back water runoff from an abandoned dump.

This particular site was for two years a commercial dump for waste from Exxon, Gulf, Union Oil Co. of California and more than a dozen others. Neighbors also tell of illegal "midnight dumping." Some of the neighbors have sued known users of the pit, charging them with contaminating their water wells; the companies are contesting the suit.

The EPA has found 37 toxic substances in and around the 17-acre site. Lead has been found in water from Mr. Sellers' well. Although no link has been shown, Mr. Sellers' wife suffers from memory loss, one of his children has kidney failure and his other two children have learning disabilities. Some other neighborhood residents complain of chronic diarrhea or nausea and of pets and livestock dying mysteriously.

Dr. Buller's Story

Small-town doctors often are left to grapple with such problems on their own. In Opelousas, La., surgeon Daniel Buller figures he treats enough cases of colon cancer to keep 10 surgeons busy.

Dr. Buller also says he diagnoses or treats an average of one case of pancreatic cancer a week. In fact, the densest cluster of pancreatic cancer deaths in the U.S. is found in South Louisiana. To use the figure for white residents, the death rate from this cause rose 57% in the area between the early 1960s and the late 1970s.

Dr. Buller worries about contamination of aquifers, which supply 85% of the state's public water systems. "It's a very frightening situation," he says. "You can envision generations being exposed to carcinogens that can never be removed." But instead of warning the public, he contends, authorities often try to cover a problem up.

If so, they have precedent. Twenty-five years ago there was an underground oil-well explosion near Gueydan, La., and a local health-department employee told his bosses in Baton Rouge that ground water might be tainted for miles around. Their reply instructed him "not to make any public statement on this situation."

Port Barre's Water

A more recent case concerns the town of Port Barre, La., which lies next to a petroleum wasteland of stinking disposal pits, injection wells and rusting separation tanks. The state health department found the town's water supply to contain more salt than the EPA's recommended limits and 14 times as much as the level the EPA considers optimum. The water tests also showed nearly twice the level of dissolved solids that federal standards permit. The state health department didn't announce the results, however; officials note that they aren't required to do so.

Nor did Porte Barre officials tell residents what the tests had found. "The decision was made to keep people in the dark," says Ray Thibodeaux, a former town councilman who attended an unannounced council meeting in November 1983 when the test results were discussed. The explanation offered by Port Barre's mayor, Roy Council, is that everyone knew the water was bad anyway.

In Kaplan, the cancer-ridden area where Jerome Vincent lives, the state health department took a very different course. Tests by Wilma Subra, who runs a local water lab, found zinc or barium or both in 81 of 84 water wells. But most residents of the heavy oil-and-gas-producing area continue to drink the water anyway. One reason is that the state health department sent form letters to many of them stating that "your water is of satisfactory chemical quality for domestic use."

The department says it based the letter on its own water tests showing contaminants to be within acceptable levels. In a few cases where the tests found barium above the safety level, later tests showed lower results, the agency says.

Limited Budget

South Louisiana's health problems have put the health department somewhat on the defense, but officials say they do all they can within the budget the legislature gives them. In addition, says Sue Ellen Lewis, the head of the agency's public-affairs office, "you have to understand who calls the tune." Asked specifically about agency handling of water-contamination problems, she says, "You'll have to pose that question to someone else. The guys' name is Edwin Edwards."

For his part, Gov. Edwards attributes South Louisiana's high cancer rate to its Cajun residents' liking for fatty foods and cigarettes. The state's chemical industry, which sponsored a study of Cajuns' life styles, blames their smoking, drinking and low dietary fiber for the region's cancer levels. The American Petroleum Institute's medical director, Neil Weaver, says he isn't aware of any epidemiological studies focusing on health and oil-field waste.

A few Louisianians are so desperate for answers they have undertaken crude studies of their own. Evelyn Allison, an organist at a retirement home in Lafayette, has traipsed through South Louisiana communities scrawling blots and stick figures on maps to mark clusters of cancer cases. "Nobody else is doing anything," she says.

But doing something isn't easy. When the state planning office assigned Lawrence Lynch to edit the 1979 "State of the State" report, he included cancer-rate data and a warning about "serious problems of air, water and ground pollution." His bosses in Baton Rouge relieved him of his duties and never published the report.

Petrochemical Waste

Pollution concerns in the coastal marshes of Louisiana focus on oil-field pits and dumps, but there is another major source of oil-industry waste in the state—the large refineries and petrochemical plants. Much of their waste now is safely sunk in so-called injection wells, but not all of it.

In the 1960s and early 1970s, a commercial disposer dumped wastes from Exxon, Shell Oil and eight other big chemical producers into a giant pit below a bluff in the Mississippi River flood plain above Baton Rouge. Some of it migrated into a marsh called Devil's Swamp, now an expanse of dead cypress trees. Flood waters commonly cover it in the spring.

The companies' wastes are no longer dumped at the pit. In response to a suit by the EPA, they have agreed to excavate the soil there to a depth of one foot. A 1982 consultant's study, however, said that a reddish sludge had saturated some soil as deep as 25 feet.

Handling of chemical waste, which the EPA regulates as "hazardous," has improved in Louisiana, but the agency nevertheless says that 40% of the state's major hazardous-waste disposal facilities violate federal regulations. This is roughly double the average for all states. In a June letter to the state, the EPA said it had found several deficiencies in ground-water monitoring at three contaminated disposal sites in Louisiana, run by Rollins Environmental Services Inc., Tenneco Inc. and Martin Marietta Corp. All deny such shortcomings and say they are cleaning up all contaminants.

In addition, a third of the big refining, processing and industrial plants that generate wastes classed as hazardous have advised the state of possible ground-water contamination at one or more of their installations. For example, PPG Industries Inc. found that toxic wastes buried at its chemical plant one to two decades ago were migrating underground and entering a ship channel at a rate of 10 pounds a day. It says the material hasn't endangered any drinking water.

The Mississippi's Water

Even injection of chemical waste into dry wells isn't foolproof. One injection well, used by Tenneco for refinery wastes, has leaked contaminants into the underground sands from which New Orleans would draw an emergency drinking-water supply, according to reports by a Tenneco consultant and the state. Tenneco says that the well only contaminated a shallower aquifier and that a cleanup program is under way.

New Orleans's main source of water, the Mississippi River, is a dumping place for Tenneco and more than 100 other refiners, sewage plants, petrochemical producers and other industrial operators in Louisiana. It isn't known whether the river dumping is related, but a National Cancer Institute study in the mid-1970s showed New Orleans' cancer death rate to be 21% above the U.S. average and the highest among 11 metropolitan areas surveyed. More recently, state water specialists began finding evidence of carcinogens in the river water.

Now Louisiana is asking the EPA to permit four fertilizer plants to dump into the Mississippi each year millions of tons of gypsum waste, containing uranium, radium and other toxic materials. An EPA staff memo to Administrator William Ruckelshaus recommends approval even though the dumping might cause the river, at low flow, to violate the federal government's water-quality standards.

A few years ago, a company called IT Corp. proposed to build in Louisiana the world's largest disposal plant for hazardous waste. It got $375,598 from the state to do a feasibility study and paid $53,836 to a consultant to help out. The consultant, however, happened to be working for the state government's own hazardous-waste division on the same project.

The Louisiana Ethics Commission last year fined the company $10,000 over the affair, which it declared had "led to actual illgotten gain by IT Corp. and others" and had "clouded the integrity of the state's hazardous-waste management program." IT, denying any conflict of interest, is appealing the commission's ruling. An IT official says, "We feel very genuinely clean and honest."

Lake Charles Citizens Try to Find Out Whether It's Safe to Go Near the Water

By a WALL STREET JOURNAL *Staff Reporter*

LAKE CHARLES, La. —There is a growing concern among residents of this refining and petrochemical center that years of industrial activity have left their water unsafe to drink.

Doctors found benzene in the bloodstreams of Mitch Martin's five children and chromium in a node taken from one youngster's neck. Shirley Goldsmith, the head of a citizens' group doing battle with chemical-waste disposers, says that on her block alone seven people have contracted cancer. "Everybody's scared to death," she says.

Her group's efforts to find out what is in the water have been frustrating. The local water board says the water is safe, but not everyone trusts the water board's word. For one thing, its chairman recently skipped town amid allegations of financial irregularities involving himself and the board. For another, the tests it cited were done with equipment lent by a local waste hauler.

The citizens' group sent water samples to an out-of-state laboratory, which reported finding dangerous concentrations of certain heavy metals. But a month later the lab disavowed its findings and said the water was fine. The firm, CompuChem Inc. of Research Triangle Park, N.C., declines to explain its turnabout except to say it decided to retest the water.

The citizen group's chief concern has been Browning-Ferris Industries Inc.'s huge waste-disposal site north of town. The company and the state have found contaminants 50 feet beneath it. Paul Jones, a retired official of the U.S. Geological Survey who has studied the site, believes toxic materials have reached a major aquifer 200 feet down. "If there ever was a clear-cut case of ground-water contamination by hazardous waste, this is it," he says.

A study conducted for the state last year concluded that Browning-Ferris should have one of its hazardous-waste disposal permits at the site rejected on grounds of unsafe conditions. The study was done by Eugene Senat, then a Louisiana State University graduate student in environmental studies. Browning-Ferris filed a $3.2 million libel suit against the researcher.

The company has since dropped the suit, explaining that its reliability as a waste disposer was publicly affirmed in a separate case. Browning-Ferris maintains that all its Louisiana disposal facilities are safe, including its abandoned sites. For several years, however, cleanup orders have been outstanding for some of the abandoned sites, including two that are near waterways and one that is 1,000 feet from a geological fault.

Browning-Ferris, which is Louisiana's largest commercial disposer of hazardous waste, has long done business with associates of Gov. Edwin Edwards. In 1972 it bought a disposal firm of which Gov. Edwards's first gubernatorial campaign manager was a director. In 1978 it bought a disposal site from a company for which one of the governor's brothers had been a lawyer and the other brother had been a real estate agent.

Currently, a public-relations firm owned by two former members of the Edwards administration acts as Browning-Ferris's governmental-affairs consultant in Louisiana. Its former registered lobbyist is a longtime friend and political associate of the governor. The company and the governor don't see any significance in such links.

APPENDIX K

ALASKA ENVIRONMENTAL INDUSTRIES, INC.

Industrial Vacuum Pumping	Formerly Mar Enterprises	Power Washing
Waste Disposal	Box 1660 • Soldotna, Alaska 99669	Roustabouts
Tank Cleaning		Power Washer &
Tank Liners	Bus: 283-5050 or 283-4906	Steam Cleaner Sales,
Oil & Chemical Spill Cleanup	Res: 262-4798	Service & Rentals

To __________, ______________________________

______________ Alaska Department of Environmental Conservation

To __________________, Drilling Superintendent, Union Oil Company of California and

To the People of the State of Alaska,

__________, I am a little tired of the two of you insulting the intelligence of the People of the State of Alaska by breaking the law and bending the rules to suit your own purposes.

Since you both seem to have a problem either in reading or interpretation, let me bring the rules as they are written, to your attention and to the attention of the People of the State of Alaska. The People have a right to know where they do or do not stand, and believe me, the People Do Stand.

Alaska Statutes the Law of our State, which by the way can be changed only by the Alaska State Legislature and not by Bob Martin, Bill Lamoureaux and or Bob Flint.

Title 46 Water, Air, Energy, and Environmental Conservation

Article 1 Declaration of Policy

Sec. 46.03.010A. It is the policy of the state to conserve, improve and protect its natural resources and environment and control water, land and air pollution, in order to enhance the health, safety and welfare of the people of the state and their overall economic and social well-being.

It is the policy of the state to improve and coordinate the environmental plans, functions, powers and programs of the state, in cooperation with the federal government, regions, local governments, other public and private organizations and concerned individuals, and to develop and manage the basic resources of water, land and air to the end that the state may fulfill its responsibility as trustee of the environment for the present and future generations.

__________, the Statutes say for the People of the State and for Present and Future Generations. They do not say for Oil Companies.

Now I know Joe, that oil represents a major slice of our economy, but I also know that our number one resource is our people.

I know too that Oil & Gas are a nonrenewable resource and that after Union Oil is gone that the People will still be here. Lets leave these People with a Clean & Green State and a good taste in their mouth.

____________ lets clean up the Union Oil Gas Field, so that we can start again to see feathers out there, on the birds and not just on your cowboy hat. The marsh is not too far gone. It can be restored. _________, if you allow either of these Union Oil disposal permits to be issued you will have violated the Public Trust, and you and the DEC will be hit with the biggest law suit in the history of the State of Alaska, one that will reverberate around the world. Let me assure you that I will not be the only plaintiff in that lawsuit. It will be the people of the State of Alaska vs. _________, the Alaska Department of Environmental Conservation and Union Oil.

To go on a bit with the Alaska Statutes, the law of our state, let me refer you to:

Section 46.03.110 Waste disposal permit procedure. Application for permit shall be made at least 60 days before commencement of a proposed discharge.

Now ____________, I suppose you can read that any way you want to but I know how I read it, and I hope your favorite friendly vacuum trucker reads it before he blows off his next load.

For the record the Alaska Statutes can not be ammended, changed or altered by the DEC or by Union Oil.

After the Alaska Statutes we next run up against the:

> Solid Waste Management Regulations 18AAC60
> State of Alaska Department of Environmental Conservation 1983, Effective October 1984

These regulations replaced the original State Solid

APPENDIX K

Waste regulations which were in effect from July 1974 to October 1984.

Under 18AAC60.075

1. A minimum vertical separation of six feet is kept between the waste application surface and the ground water table impossible in the Cannery Loop Area and impossible in the Gas Field.

2. A minimum horizontal separation of 100 feet is kept between the disposal area and surface waters, impossible in the Gas Field, impossible at Cannery Loop 3, questionable at Cannery Loop 1.

3. Surface runoff does not enter or leave the disposal area. It is entering and leaving now in both areas and has been for years.

Under 18AAC(3) Drill Mud disposal of on land is confined in a pit, trench, or diked area that prevents erosion, or mass flow of the mud. Authority AS 46.03.020(10)

Bentonite erodes every time a vacuum truck discharges its load, or it rains hard, there is erosion between the Bentonite and the native soil.

In areas of loose porous soil such as Cannery Loop and the Gas Field Bentonite will probably, not work, but even if it did you are in violation of the Regulations and the Statutes.

Under 18AAC60.200 A permit is required by any person who constructs, modifies, or operates a solid waste site. And he shall do so in accordance with a waste disposal permit issued to that person by the department.

A permit is not required for a single family or duplex residence or a farm where solid waste is generated and disposed of on the premises.

I haven't seen any cows, pigs, or chickens in the Union Oil Gas Field so I don't think the Union Oil Gas Field meets these criteria, but they have been disposing of mud and not storing for years regardless of what stories they tell to the contrary.

Under 18AAC 60.210 B The Application must show how the proposed activity will comply with this Chapter and must include two copies of a certificate of compliance with local ordinances? Union Oil is the City of Kenai happy with the way you have complied with local ordinances?

APPENDIX K

A written statement from the landowner, if a person other than the applicant, giving the applicant permission to conduct the proposed activity. Union Oil do you have this, and if the landowner knows how unhappy the people are, would that landowner then grant you the required permission to continue adding to the Peoples Unhappiness?

Also __________ how can you show that the proposed activity will comply with this Chapter when:

A. You can't keep six vertical feet between the water table and the bottom of your mud pit.

B. You can't keep 100 feet between your disposal area and the surface waters.

C. You as yet have not stopped surface runoff from entering or leaving the disposal area. I'm not saying you can't, but you haven't done it yet.

I have not as yet gone into any detail on Regulations or Statutes for the Federal Government, the Kenai Peninsula Borough and or the City of Kenai, I will go into these if and when I feel that it is necessary.

At the present time this issue is being investigated and or being looked into by:

A. Jack Chenoweth, the ombudsman in Juneau

B. Alaska State Department of Fish and Game, Habitat Division

C. Criminal Investigation Division of the E.P.A., Region X

The issue has been dumped into the laps of the top 100 newspapers in the United States by circulation. The issue has now been dumped into the lap of the Associated Press. Before long we will be on TV, and I haven't even begun to fight yet.

You are breaking the Statutes and the Regulations. You are going against the will of the People of the State of Alaska. You are violating the Public Trust.

If you continue __________ in this blind, senseless, illegal direction you are now headed in, you may find that you embarrass Governor Sheffield sufficiently so that he asks for your resignation. Politics, money, and lobbying are fine for a while, but there comes a time when its wiser to retreat than it is to advance.

__________ its been good talking to you. I knew

you'd understand. I'll see you both at the packed Public Hearing in Kenai tonight.

To the People on the Kenai Peninsula, thanks again for listening and I'll see you in Kenai at 210 Fidalgo St. tonight. Thanks for all your letters, and phone calls, and your offers of support.

Sincerely,
Dave Brown
Keeping Alaska Clean,
For Your Kids and Mine

P.S.: 18AAC 60.910
(8) "Drill Muds" means the fluids composed of clays, oil, water, and chemical additives used by oil well and other deep drilling operations to lubricate the drilling bit, maintain hydrostatic pressure control in the well, remove drill cuttings from the well and stabilize the walls of the well during drilling.
(30) "Solid Waste" means garbage, refuse, sludge, and other discarded material, including solid, liquid, semi-solid or contained gaseous material resulting from industrial, commercial and agricultural operations and from community activities.

Thanks again.

APPENDIX L

PENINSULA CLARION

11/27/84

Landowner opposition hits Union

Drilling mud disposal unlikely without consent

By JAN MIRELES, Staff Writer

A hammer may be chipping away at Union Oil Company's bid to dispose of drilling mud in the Kenai wetlands.

Early this fall, Union filed a two-inch thick permit application with the state Department of Environmental Conservation proposing to dispose of drilling mud at seven sites. Five of the sites are located in the Kenai Gas Fields off Kalifornsky Beach Road. The other two are Cannery Loop Units 1 and 3. Both are inside the Kenai city limit.

While Union holds the mineral lease on the gas field sites, it does not own the land. According to state regulations, Union's permit cannot be granted until the firm produces written statements from each landowner granting permission for the drilling mud disposal.

Union has produced no such letters, say state officials. Without the letters Union will have to forego the proposed disposal sites, said Bob Cannone, district manager for the DEC.

"There's no question that if the landowners don't grant permission then that's that. They've got the final hammer," he said.

The hammer is held by the Salamatof Native Association, Leonard A. Keener, Doris M. Lashley, Cook Inlet Region Inc. and the Kenai Peninsula Borough. All own proposed disposal sites in the Kenai Gas Fields. Keener shares ownership of one of the sites with the borough.

All have questioned Union's application, saying the oil firm has yet to approach them for permission to dispose of drilling mud on their land. Only the borough has said it would consider giving approval.

Six gas wells have been drilled at site 14-6 in the Kenai Gas Fields. The land is owned by the Salamatof Native Association. Andy Johnson, the group's president, said the association would deny any request from Union to use the land as a disposal site.

Johnson said the agreement with Union requires the firm to return the land to its original condition when drilling operations are completed. Drilling mud disposal is not part of the bargain, he said. He also said telegrams were sent to the DEC and to the governor's office protesting the proposed site.

Keener owns a portion of site 34-31. The borough owns the rest. He fears the drilling mud may contaminate local groundwater.

"They haven't approached me and I wouldn't sign it (a letter of permission) if they did. I'd feel bad if they polluted the water," he said.

APPENDIX L

Lashley is adamant about her opposition to drilling mud disposal on her land. Three gas wells have been drilled at site 33-30, but Lashley said she has made no agreement with Union to allow permanent disposal of drilling mud.

"I am very unhappy with Union Oil," she said. "I would never, never grant permission."

Equally vocal is Cook Inlet Region Inc. CIRI Vice President Frank Klett is irritated that Union has not contacted his group about the permit proposal. Klett said it is unlikely permission will be granted.

"We're not in the business of disposing of drilling mud," Klett said. "And we don't intend to be."

While the borough has not been contacted for permission to dispose of drilling mud on its land, Borough Mayor Stan Thompson said there is no ordinance prohibiting the borough from allowing disposal. Thompson did say Union would have to meet all DEC requirements and that the borough would consider public opposition to the proposal before making a decision.

Union's application has met with widespread opposition from the public as well.

Residents living near the gas fields fear the drilling mud may contaminate drinking wells. Environmentalists and conservationists say the proposed disposal sites are within the Kenai wetlands and pose a danger to wildlife.

Dave Brown, operator of the Sterling Special Solid Waste Site, claims the application is illegal because his is the only site licensed to handle drilling mud disposal. Union's refusal to use the Sterling facility has placed Brown's business in peril, said Brown. Brown has filed for partial bankruptcy—asking the court for time to reorganize his business.

Also in question are Union's proposed disposal sites Cannery Loop Units 1 and 3. The units are located inside Kenai city limit and are subject to a city ordinance addressing management and construction of solid waste sites.

The Clarion reported last week that the ordinance may prohibit Union from disposing of drilling mud at those sites. The ordinance grants the borough complete authority in contracting for solid waste sites. Kenai City Manager Bill Brighton said the only way Union could dispose of drilling mud inside the city would be to get approval from the borough.

While Thompson said the borough might agree to such an arrangement, he said it would depend upon the city's preferences.

"If the city opposed them then I'm sure the borough would, too," he said.

In October, Kenai Mayor Tom Wagoner sent a letter to the DEC and the governor requesting the state deny Union permission to dispose of drilling mud at either site.

Despite the strong opposition, Union's Alaska representative, Bob Anderson, said the firm still hopes to have its permit approved.

Anderson denied trying to side-step regulations calling for landowners permission to dispose of drilling mud. "I don't think that's the case at all. Obviously not," he said.

He said the people putting together the application must have believed the information was not yet needed by the state.

"I think we're going to have to wait and see," he said. "And if additional steps are needed, obviously we'll do it. Union is hopeful that they'll approve the permit and we can get back to business."

Drill mud studied by FBI

The FBI is contemplating launching an investigation into the disposal of drilling mud on the Kenai Peninsula by Union Oil Company, according to an Anchorage agent.

"The Environmental Protection Agency has brought questions to our attention," said Lawrence Nelson, agent in charge of the Alaska FBI. "They would have no objection to our investigating the issue."

Union Oil has filed a permit request with the state Department of Environmental Conservation to dispose of more than 126,000 barrels of drilling mud in and around Kenai. The request has met with public opposition. So far the state has denied use of four disposal sites. State DEC officials expect to make a final decision on the permit application by March.

Nelson said the EPA normally would have primary jurisdiction over an investigation of drilling mud disposal. Since the EPA has indicated it does not have the resources to get involved, the FBI may take on the case, Nelson said.

The FBI, however, is uncertain about what approach should be taken. Nelson said the agency has never investigated an environmental issue in Alaska.

"When you've not done something before, you don't enter into it willy nilly until you're sure that there's a reason to get into it," Nelson said.

Nelson said he would try to determine the possibility of federal violations before making a decision.

"Within the next week I'll attempt to assess the nature of the problem and what role we should play," he said.

Opinion

The muddy issue

Union Oil appears to have backed unwittingly into a blind corner in the matter of drilling mud disposal on the Kenai Peninsula.

Despite the staffing and resources that large oil companies can summon, and despite the filing of a purported two-inch document of application with the state, it appears that Union overlooked provision No. 7 [technically: 18

APPENDIX L

AAC 60.210. APPLICATION FOR PERMIT . . . (b)(7)] of the application guidelines: You have to get the property owners' permission.

Whether Union officials actually overlooked the provision or were overconfident about their ability to get the permit first and the permission later, is a matter of conjecture.

After public hearings and weeks of scrutiny by this newspaper, one question looms above all others: Why should Union be allowed to do anything other than provide for disposal in the property provided location?

That location happens to be the special waste site in Sterling.

Does anyone believe that Union Oil cannot afford to do this? Of course not.

Then the logical conclusion seems clear. If it's in the public interest to dispose of a questionable substance according to a certain legally established procedure, and especially if the operator in question can afford the procedure, then it ought simply to be required.

There is the prospect of a curious side benefit in the case of Union and the Sterling special waste site. If the site operator, Dave Brown, has his calculations right, Union could fill the site up in a year or so.

Thus, filling the Sterling Special Waste Site would actually prove to be the end of the long-standing controversy about its location. If it were closed, covered over and abandoned, it would accomplish, in a sense, what a number of Sterling residents have been unable to bring about through other recourse: the closure of the special dump.

It has become clear in recent weeks that a substantial number of Beaver Creek area and Kalifornsky Beach Road area residents are concerned about drilling mud disposals; and that the private owners of Kenai area lands where Union would like to have disposal permits are opposed to the idea.

The borough has ownership interest in some of the land involved, but it is hard to imagine anyone arguing that the borough should accede to Union's wishes. The public owns the borough land, and the public interest is clearly served by simply asking Union to do what it's supposed to do.

There appears to be no justification for the borough helping Union to get its way.

There appears to be no justification for the state's Department of Environmental Conservation granting permits as Union has requested.

There appears to be every good reason for Union Oil to demonstrate public responsibility and assure that all of its drilling muds are handled according to law.

Union Oil may not be selling gas on the peninsula, but one of its divisions, Union Chemical, operates a plant in Nikiski. The good community relations established over years by Union Chemical could be erased by irresponsible action from Union Oil.

APPENDIX L

Kenai Natives join drill mud protesters

The public comment period for Union Oil Co.'s drilling mud disposal application ended quietly Tuesday. But public concern over Union's request to dispose of drilling mud in the Kenai wetlands has not abated, said a state official.

"There are a lot of people on the Kenai concerned about this," said Bob Flint, regional director for the state Department of Environmental Conservation.

"That's not always true of other areas," he added.

Tuesday the Kenai Native Association joined the list of groups opposing the disposal request. The association sent telegrams to the governor's office and to the DEC asking that the permit be denied, according to an association worker.

Originally the public comment period was to have ended in October. The deadline was extended, however, following requests from concerned residents. The state also held a hearing in Kenai on Oct. 29 and another in Anchorage on Oct. 30. During the hearing state and Union officials explained both the permit application and the review process.

Flint estimated more than 25 people addressed the department during the hearing. A crowd of about 125 people attended the 3½-hour meeting. Added to other letters received before or after the hearing, Flint said, more than 50 people have voiced concern about the safety of drilling mud disposal.

Bob Martin, special assistant to the state DEC commissioner, said he is aware of only two or three letters received since the hearing. He would not say when the state might reach a decision concerning Union's permit application.

"We are still reviewing the permit information," Martin said.

The state also is working on developing a set of guidelines for drilling mud disposal, he said. Despite urgings from environmentalists to postpone any permit decisions until a set of guidelines is in place, Martin said there has been "no decision to act one way or the other."

"It's probable in fact that we will get back to Union Oil with a request for more information," Martin said. "And it's also possible that we'll delay a decision until we do develop a more comprehensive policy on the disposal of drill muds."

Martin said the state will keep the public informed on its progress concerning the department's review findings.

"It's obvious that people are concerned and interested," he said. "We're not just going to sneak a decision by."

APPENDIX L

Denial of permit to bury drilling mud urged

From our Peninsula Bureau

KENAI—Kenai Mayor Tom Wagoner has asked the state to deny a permit application that would allow Union Oil to bury 126,000 barrels of used drilling mud in and around his city.

Union wants to dispose of the mud in seven existing reserve pits in the Kenai and Cannery Loop gas fields. Some of the pits are located in wetlands areas and some are located near residential subdivisions.

In a letter to Gov. Bill Sheffield, Wagoner opposed use of existing pits for drilling mud disposal. He requested, instead, that the company be required to "dispose of all drilling muds at one site within the Kenai gas field."

This would "ensure that a new pit is constructed to exact specifications" determined by the Alaska Department of Environmental Conservation.

Finally, Wagoner asked the state to force the company to clean up and remove all drilling muds remaining in the Cannery Loop field, which is located entirely within the city.

Wagoner's letter could signal a change in attitude on the part of local officials towards the oil industry. Two years ago, at Union's request, the city council modified the city drilling ordinance.

The first change allowed Union to substitute earthen reserve pits for steel tanks. A second made it retroactively legal for Union to bury used muds in the city.

DEC will not decide on Union's permit application until later this year. The agency will accept written comment on the request until Nov. 13.

Possible mud disposal site rapped by Fish and Game

By JAN MIRELES, *Staff Writer*

The state Department of Environmental Conservation may issue Union Oil Co. a permit to dispose of drilling mud on a Kenai Gas Field site, despite opposing recommendations from another state agency.

In a November 23, memorandum to the DEC, the state Department of Fish and Game urged officials to require Union to clean up four proposed drilling mud disposal sites.

The letter asked that Union "be required to backhaul all muds and cuttings from Sites CLU No. 3, 14-6, 14-18 and 43-32 to an approved upland site." The move is needed because the pits are contained within gravel pads in wetlands areas, Fish and Game officials said.

Last September Union filed a permit application with the DEC asking permission to dispose of 126,000 barrels of drilling mud at seven sites; five on the Kenai Gas Field and two within the Kenai city limit. More than 100 local residents later opposed the proposal at a public hearing held in October.

Then on Nov. 27, the DEC announced a preliminary decision to deny disposal at four of the sites. The sites included CLU No. 1 and CLU No. 3, both located in Kenai; and gas field Sites 41-18, and 14-6.

Department officials said the decision was "based strictly on water tables." The department also announced that it hoped to make a final decision regarding all the proposed sites by March.

The list of denied sites did not include gas field site 43-32. According to the Fish and Game report, however, Site 43-32 already has contaminated wetlands. The report reads in part:

"The ADF&G is especially concerned that the muds from Site 43-32 be removed. This pit was built without a bentonite liner, and percolation of fluids from under the toe of the dike has been documented. Adjacent wetlands have been contaminated."

The report said the muds buried at the site contain chemicals known to cause environmental damage and that the pit is leaking.

All of the pits, said the report, contained barite, lignosulfonate and concentrations of chromium, lead, barium, gas and diesel oil. "Because these materials are present in the pits that Union proposes to bury, any leakage will probably result in degradation of the surrounding environment," said the report.

In summary, the report warns that drill mud leakage would damage the Kenai wetlands. "Vegetation may be killed directly by salts or by high heavy metal concentrations. Increased heavy metal concentrations in plants may increase heavy metal concentrations in herbivores. Aquatic insect populations may also be reduced, and nesting populations of birds that depend on these insects may also be reduced. Overall," said the report, "drilling mud leakage would result in environmental degradation of the area surrounding the buried pits."

Bob Flint, a regional program coordinator with the DEC, said the Fish and Game report did not affect his department's decision.

"It didn't have anything to do with our response. Fish and Game was a little late in getting to us," he said.

The report, Flint said, did not reach DEC officials until Nov. 27, the day the preliminary decision was announced. Flint said it is unlikely the department will address Fish and Game recommendations until their March decision date.

"What the department doesn't want to do is, based on some partial information, make somebody do something for no reason," Flint said.

Flint refuted the urgency of the Fish and Game letter and said the DEC was researching possible problems with the sites itself.

APPENDIX L

"We're not holding off any decision," he said. "We are developing one document with questions for further consideration of the sites."

The questions, he said, will then be given to Union Oil to answer. Flint said he expects the questions will be prepared within the next two weeks.

APPENDIX M

Drilling mud unearthed

Waste site operator excavates Beaver Loop site

By JAN MIRELES, *Staff Writer*

A local waste site operator who claims Union Oil Company illegally disposed of drilling mud on the Kenai Peninsula has uncovered the mud at one of the firm's abandoned sites.

Dave Brown, operator of the Sterling Special Waste Site, hired a backhoe to dig up a portion of Union's Cannery Loop Unit Two located off Beaver Loop in Kenai. The backhoe, digging a hole measuring approximately 30-feet long and 13-feet deep, uncovered drilling mud throughout.

Union spokesman Mark Eversole discounted the find, however, and said the company has a permit from the Alaska Corps of Engineers allowing disposal of drilling mud solids at the site.

"We've never denied that there was drilling mud there," Eversole said. "That was spelled out in the application. We stated there would be a pit, that we would pump off the pumpable liquids and bury the solids."

Pat Richardson, assistant public information officer for the corps, denied any knowledge of a Union permit. According to Richardson, the corps grants permits only for clean fill material.

"We don't regulate drilling mud," she said. "Drilling mud would have things in it that we wouldn't allow under our permit."

Although the state Department of Environmental Conservation is considering a Union permit application to dispose of drilling mud at seven sites, Cannery Loop Unit Two is not part of the proposal. The Alaska Oil and Gas Conservation Commission approved abandonment of the site in 1982.

C.V. Chatterton, chairman of the commission, said a site is approved for abandonment when the surface is clear and free of all trash and debris. Chatterton said the AOGCC is not responsible for what might be buried underneath abandoned drill sites.

"I don't know who would be," he said. "That's a good question. As far as this agency is concerned the site was cleaned up. We don't worry about waste disposal. That's the turf of another state agency."

Local DEC officer Bob Cannone said he was not certain which agency bears responsibility for making sure the site was cleaned. He said he would be interested in seeing proof that drilling mud was found at the site.

"Whoever issues the permit is responsible," he said. "It would be their (AOGCC) responsibility to police the area."

In most instances the oil and gas commission issues drilling permits to oil companies and then notifies the DEC, Chatterton said. The DEC is notified, he said, so they will be aware firms may be seeking additional permits for related drilling activities.

The DEC is developing guidelines for drilling mud disposal and officials have said they expect to unveil a policy by April. Brown has argued that Union has violated solid waste disposal laws and should be forced to remove the drilling mud immediately. Brown operates the only licensed disposal site for drilling mud on the peninsula.

Brown originally estimated that 35,000 barrels of drilling mud had been buried illegally at the Cannery Loop Unit Two site. The original dimensions of the pit were 400 feet long by 80 feet wide by 15 feet deep. When a backhoe operator started digging the hole Tuesday he reached the water table at approximately 13 feet, Brown said.

Also found in the pit were plastic sheeting and bits of discarded lumber. The drilling mud, said Brown, already may have contaminated the surrounding water table.

Landowner Michael Pelch of Oregon gave Brown written permission to dig up the pit. Pelch has told the Clarion that he does not want drilling mud disposed of on his property.

Representatives from the Environmental Protection Agency and the Alaska Corps of Engineers were expected to visit the site this morning, Brown said.

APPENDIX N

ALASKA ENVIRONMENTAL INDUSTRIES, INC.
Formerly Mar Enterprises
Box 1660 Soldotna, Alaska 99669
Bus: 283-5050 or 283-4906
Res: 262-4798

February 8, 1985

To the People,

On the front page of the Peninsula Clarion Jan. 23 was the article about drilling mud unearthed, together with my picture standing in a hole at Union Oil's abandoned Cannery Loop No. 2 site on Beaver Loop Road.

I'd like to clear up all the misconceptions and deception surrounding Cannery Loop No. 2 as well as possible misquotes by the press in the Anchorage papers.

To begin with Michael Pelch the landowner in fact asked me to dig up Cannery Loop No. 2. He also ordered the D.E.C. and Union Oil to remove immediately any and all mud that was found.

D.E.C. says they wanted to know the truth about Cannery Loop No. 2. Nonsense, I told D.E.C. the truth about Cannery Loop No. 2 four years ago. Howard Keiser who worked 17 years with D.E.C. told Union Oil to move the mud. Union Oil ignored Howard. Then Howard retired and the issue died.

The D.E.C. is computerized and not programmed to handle the truth.

I have discussed Cannery Loop No. 2 at least two dozen times with Bob Martin, Bill Lameroux, Bob Flint and John Dough. My words were wasted on deaf ears. They don't care about any environmental problems at all, unless they are embarrassed by them.

Union Oil says they had a permit from the Corps of Engineers, and in the Anchorage papers Union said they weren't worried because of the pits liner.

Nonsense in both cases. D.E.C. is the only agency that can in the State of Alaska issue a permit to dispose of drilling mud. Union Oil knows this, and excess use of words in a Corps permit does not remove the liability for the deliberate violation.

As to Unions usual ploy, "it is merely being stored" , nonsense again. You don't store drilling mud with visqueen, scrap lumber, rocks & dirt for four years, and you don't store drilling mud under six feet of dirt.

The biggest joke of all is the pits liner. There is no liner in that pit, and there never was.

A thin bentonite liner was sprayed on the side walls, 1/8″ to 1/4″ thick. We were asked to spray water over the bentonite which we did. The liner was as worthless as cheesecloth, and you could see gravel through it everywhere you looked.

APPENDIX N

There was no liner in the bottom of the pit as there was one foot of ground water in the pit bottom. The pit was and still is in the water table.

The only conceivable reason for the worthless liner was to give Union or __________ some back up paper work, so they'd look good on paper if there ever was a problem.

The drilling mud today at Cannery Loop No. 2 is 8 feet deep, 400 feet long and 80 feet wide.

There lies buried illegally at Cannery Loop No. 2, 35,000 barrels of drilling mud, not stored, buried with no liner, directly in the water table and without a D.E.C. permit.

In the oil industry one barrel is forty two gallons.

So including bentonite, diesel, caustic soda, and all the other various and sundry additives that Union uses in the mud, there is a total of 1,470,000 buried gallons at this one location. Union Oil has many similar locations, unlined and leaking, all illegal, most directly in the water table.

Union Oil and the D.E.C. are playing Russian Roulette with your health, your welfare, and your peace of mind, if you live in one of the effected areas.

__________ is starting to sound like Richard Nixon, but maybe his mistake lies in accepting as the truth, that which __________ tells him.

Mike Pelch the landowner is subdividing and can't even get the mud removed. How is he going to sell his lots?

And in Juneau they are worrying about Socialism on the part of Bill Ross our new candidate for commissioner?

WHAT HAVE WE GOT RIGHT NOW?

Does your drinking water taste or smell different lately?

Don't worry folks, our D.E.C. is on the job. The original set of solid waste and drilling mud regulations were written in July 1973. Union Oil ignored those and D.E.C. didn't enforce them. On October 1, 1984 the regulations were rewritten to protect you and I. Again Union ignored those and D.E.C. didn't enforce them.

Now those regulations are being rewritten for the third time to give Union Oil a badly needed escape hatch and to give D.E.C. a way to hide their embarrassment.

As long as our Department of Environmental Cover up is not embarrassed and Union, after all these years is finally made honest at the stroke of a pen, then why should you and I care!

Sincerely,
Dave Brown
Alaska Environmental Industries, Inc.
Keeping Alaska Clean

APPENDIX O

DEC negotiating Unocal lawsuit

POLLY CRAWFORD
Associate Editor

The state Department of Environmental Conservation and Unocal are in negotiating stages concerning the lawsuit the DEC filed against Unocal over the burial of drilling muds at Cannery Loop No. 2.

Cannery Loop No. 2, located off Beaver Loop in Kenai, was a drilling site Unocal abandoned, leaving behind drilling muds covered with dirt. Because Unocal left the waste without applying for a permit, DEC filed suit.

According to Bob Flint, regional program coordinator for solid wastes with the DEC, the court requires that the opposing parties of civil suits negotiate. He said they are currently working on a test program acceptable to both parties, to see what is really in the pit.

Drilling muds were uncovered in February of 1985 by the former operator of the now-defunct Sterling Special Waste Site, Dave Brown. Flint said the content of the muds and what else is in the pit needs to be discovered, along with the potential for leaching. How far the mud is above the ground water is also still an unknown, Flint said, although according to the U.S. Army Corps of Engineers, the site is in wetlands.

Flint said that once the testing procedures are worked out, they will have to negotiate as to whether the area could be cleaned up or buried on site. An inability of DEC and Unocal to reach an agreement will mean the court will have to ultimately settle it.

Although Cannery Loop No. 2 is the only drill mud site which has been taken to court, the other mud pits in the Cannery Loop unit and the Kenai Gas Fields are awaiting decision by a hearing officer. According to Flint, the final briefs were filed with the hearing officer in May.

Unocal applied for permits to bury the rest of the muds where they are, but DEC denied the permits, which is why the issue was taken before an adjudicatory hearing officer—one step away from a court hearing.

State says Unocal must empty waste pits in Kenai wetlands

ASSOCIATED PRESS

Unocal Corp. must empty oily waste from leaking pits in wetlands near Kenai, a state hearing officer has ruled.

The decision comes four years after the state Department of Environmental Conservation rejected Unocal's proposal to just cover up the pits. The state wants the company to remove the material, primarily drilling muds and cuttings.

Bob Anderson, who heads Unocal's Alaska's office, said the company has not decided whether to appeal the decision.

APPENDIX O

He said there are about 60,000 cubic yards of material in the seven pits. He said to remove the material and transfer it to a disposal site likely will cost millions of dollars.

The pits haven't been used in years, and the company periodically drains water that accumulates from rain and melting snow, he said.

In 1985, the DEC denied Unocal's request to leave the drilling muds in the pits, because the pits were unlined or poorly lined. Some have been leaking and the DEC has measured toxic contaminants in wetlands adjacent to some of the pits, although it can't establish the pollution came from the pits.

Two pits are within the city of Kenai and others are near residential areas.

Bill Lamoreaux, DEC's southcentral regional supervisor, said the state wants Unocal to submit a disposal plan including where the material will be taken and how it will get there.

He said it may take several years to clean out all the pits. "I'd say it would be pretty ambitious to remove all that stuff in one year," Lamoreaux said.

Kristine Benson of the Alaska Center for the Environment said it's "imperative" that DEC expedite the cleanup process. The center has been a critic of the Unocal pits, along with other Kenai pollution problems.

"We asked the department five years ago to enforce against the company," Benson said. "The fact that they chose to go through a long administrative process makes it that much more imperative that the schedule be hastened now to take care of the problem.

APPENDIX P

Poppy Lane oil waste dumpers caught in act

By POLLY CRAWFORD
Associate Editor

The fog hung mysteriously over the still, gray waters in the gravel pit, but the many tire tracks through the mud from the open gate that leads to the Kenai Gas Fields, along with the faint smell of diesel wafting through the air, were telltale signs something had happened Tuesday night at the pit at the end of West Poppy Lane.

Residents and state Department of Environmental Conservation officials agreed something was dumped. Residents have photos that the trucks came from the Kenai Gas Fields, operated in the area by Union Oil.

Both residents and DEC officials agree whatever was dumped contains some sort of hydrocarbon.

But DEC office manager Bob Cannone said he didn't know Wednesday how much contamination was there. "There was an oily smell," he said. "Whether they were trying to bury something illegally, or there was just a small contamination, I don't know."

At this point, those who do know aren't saying. A phone call to Saltz General Contractors, which residents claim did the hauling, was not returned, and a worker at the gas field said he couldn't say anything, that any information would have to come from the Anchorage office. The Anchorage office referred this reporter to Union's Los Angeles public relations officer, Mark Eversole, who said he didn't know anything about it, but would look into the matter.

The pit has a history of dumping, including drilling muds, gas well condensate, and some kind of chemical storage. Cannone said he found some substance oozing up from the ground in the vicinity of the chemical storage, but the lab was not able to identify it.

Wednesday was the deadline the DEC had given Union to respond as to its intentions concerning the pit. "We want holes dug to expose any material," Cannone said. In response to the deadline, Union has requested a meeting on Friday with the department's Anchorage office.

Many of the wastes have been in the pit for several years, and when asked why they are still allowed to be there, considering the area surrounding it is a residential neighborhood with shallow wells, Cannone said, "We've been trying to work with Union to get it removed, rather than go to the court. But the department's attitude is changing. We've gotten burned. People are taking advantage of us."

He said in the past the department has had limited access to attorneys, but that now they may be taking more companies to court.

APPENDIX P

Pollutants found near Soldotna

By RONNIE CHAPPELL
Daily News reporter

Cancer-causing chemicals have been found in the water table beneath an oil company dump site in a populated area near Soldotna, state officials said Monday.

Testing indicates that none of the private wells at nearby homes have been contaminated. But state officials did find "indications of benzene" in a barn well located north of the dump site.

Unocal Corp., owner and operator of the dump site, is trying to clean up the site by pumping large amounts of oily groundwater out of a deep pit in the northwest end of the 50-acre dump site.

State officials have said the oily substance is a liquid by-product of natural gas production, called by the oil industry a gas field condensate. The by-product, which is similar to kerosene, contains benzene, which is listed as a carcinogen by the U.S. Environmental Protection Agency.

"There are several areas where gas field condensate has been deposited in the soil," said Bruce Erickson of the Alaska Department of Environmental Conservation. Groundwater in the area is only 10 to 18 feet deep.

"The water table in the northwest corner" of the site "has been contaminated with this substance," Erickson said. "When you dig down, you can see an oily sheen floating on the surface."

The dump site is located off Poppy Lane, near Kenai Peninsula Community College.

Although state officials discovered the foul-smelling substance in the water table in mid-July, they waited until early October to order a cleanup. The delay was necessary to allow for laboratory analysis of the substance, Erickson said. Identification of the material was crucial to drafting a plan for proper cleanup and disposal of the oily liquid.

Since Thursday, Unocal has sucked more than 400 gallons of condensate out of the water table, Erickson said.

"I think (the contamination) is very localized," said Unocal spokesman Bob Anderson. "In any event, we'll do what it takes to get it out of there. We don't think there's much of a volume."

The July investigation was prompted by the complaints of area residents who discovered Unocal dumping contaminated peat at the site more than a year ago.

The review also turned up several small pockets of used drilling mud, two barrels of glycol, pipes, old well-heads and household garbage. Drilling muds can contain high concentrations of arsenic, chromium and other heavy metals.

Unocal has removed a large quantity of contaminated peat from the dump site, Anderson said. Efforts to incinerate the peat in a small asphalt batch plant were unsuccessful. The material has been bagged and is awaiting shipment to a hazardous waste disposal site in Oregon.

To date, tests of nearby residential wells have been negative, Erickson said. The water table appears to flow away from most of the homes in the area.

Test wells maintained by the oil company in and around the dump site show no sign of benzene contamination, Anderson said. Tests done in late 1985 did reveal concentrations of heavy metals which exceeded the state's safe drinking water standards.

It is too early to predict if Unocal will be able to clean up the underground chemical spill, Erickson said. It's not known how far the condensate has spread or whether it has reached a deep aquifer into which several adjacent wells are drilled.

Local residents have stopped drinking water from their wells, said Kenai attorney Peter Erhardt. Many are wondering whether they should continue bathing in it. Benzene can be absorbed through the skin.

Erhardt represents a group of adjacent landowners who filed a $3 million lawsuit against the California oil company late last year. The residents contend that materials buried at the dump site have destroyed their property values, polluted their wells and affected their health.

Residents say health declined by Poppy Lane pit

The chromium that was found in the monitoring wells at the West Poppy Lane gravel pit in June may have found its way into residential drinking water last year.

Jessica Black and Bob Sizemore, who own a house adjacent to the pit, said they experienced health problems last year sometimes associated with heavy metal poisoning.

Black said they were both experiencing a record number of headaches, and were often nauseated. "I experienced inexplicable fatigue," she said. "I was drugged out . . . sleeping 10, 12, or 14 hours a day."

She said she went for her annual exam in January and blood test results showed chromium levels in excess of twice the normal amount. Black immediately stopped drinking the water.

She found out later that the best way to measure chromium was through a special type of urine test. When Black took that test six months after she stopped drinking the water, the test revealed her chromium level was back to normal.

Black was never tested for barium and arsenic, two other heavy metals found in the monitoring wells. Her husband was not tested at all while he was drinking the water, but his symptoms did disappear after drinking the water.

Black said her health problems have disappeared except for some heart problems, which she said she was told by the state epidemiologist might be a result of barium, although she has never been tested for that.

Black said when she got the test results, "I was shocked. I was so mad and frightened at the same time. I don't want to die of cancer just because Union wants to make a profit."

Unocal, formerly known as Union Oil, the owner of the pit, has denied dumping anything in the pit which could be a source of the chromium, barium and arsenic.

Three other residents with property adjacent to the pit filed suit Friday in Kenai Superior Court against Unocal and Saltz Construction, whose trucks were seen dumping material into the pit.

The suit was filed by attorney Peter Ehrhardt and Bill Bixby on behalf of Charles and Sheila Dickson, Mark and Karen Torwick and Pete and Kris Reed.

While the plaintiffs have not experienced health problems yet, Ehrhardt said potential future problems were cited in the suit. But, he said, the main emphasis is the loss of property value.

"Who would buy that property now," Ehrhardt said. "A person's property is their life's work, and Union has destroyed it."

The suit asks for $1 million in accord with proof resulting from a trial. It also asks that Union clean up the pit and prevent any further dumping at the site.

Unfiltered tests by the Department of Environmental Conservation of the proximate 10-foot monitoring wells indicated high levels of the three metals, and while Unocal's own filtered tests showed lower levels, some of the tests were still significantly higher than the drinking water standards.

Tests last April of Black and Sizemore's well, which is less than 30 feet, indicated the levels were within drinking water standards.

Unocal spokesman Bob Anderson had also previously denied any dumping of peat contaminated with gas well condensate, a hydrocarbon, but now admits there was some dumped in the pit, and has agreed to clean it up. However, the condensate is probably not a source of the heavy metals.

A small amount of drilling muds was found in the pit, but Unocal denied any knowledge of the muds. Drilling muds often contain heavy metals.

Poppy Lane residents sue Unocal

By POLLY CRAWFORD
Associate Editor

Jessica Black and Bob Sizemore have filed a lawsuit against Unocal (Union Oil Company of California) for the loss of their property value near the West Poppy Lane gravel pit.

It's the second lawsuit to be filed against the company over the pit. The first was filed by Charles and Sheila Dickson, Karen and Mark Torwik, and Ron and Vikki Engleman.

Several weeks ago, the borough took action on several pieces of property next to the pit, placing a $1 value on them.

The complaint, filed in Kenai Superior Court this week, asks for no less than $250,000 in compensatory damages and $500,000 in punitive damages.

The complaint states that shortly after Sizemore and Black began building their home in 1984, they asked a representative from Unocal if the gravel pit contained any toxic materials, hazardous subjects, or waste materials resulting from oil exploration or development. The complaint says Unocal said there wasn't any.

The complaint further alleges that Unocal knew that was false, and in fact utilized the site as a hazardous waste disposal site. The company, at the direction of the state Department of Environmental Conservation, spent last fall and this spring attempting to get rid of gas condensate from the ground water in a portion of the site.

The complaint also states that the defendant "has allowed to be placed or deposited upon the premises owned by it or under its control matters and things which are obnoxious and offensive to the public and which produce, aggravate and cause the spread of disease and endanger the health of the community."

In a letter written to Mary Sisson, head of the Citizens Awareness Committee, Gerry Graham of Unocal stated that the gravel pit was used as a place to dump peat removed from the Kenai Gas Fields for the construction of roads and pads, but that use is not regulated. The letter also states that scrap metal, empty barrels and 7 to 10 barrels of condensate-contaminated Glycol were buried in 1971, before disposal was regulated by Alaska state law.

The letter also states, "Until the mid-1970's inactive areas of the gravel pit were utilized by the gas field operator and, without consent, the local residents as a dump for general household type trash. The trash was buried at the direction of DEC in the mid-1970's when government waste facilities were operated and a multitude of such informal locations on the Kenai Peninsula were closed."

The letter states that efforts to locate toxic or hazardous materials have been ongoing since October 1984. Yet residents took pictures of trucks dumping what appeared to be condensate-soaked peat in the pit the fall of 1985.

The letter also states that Unocal has spent close to $500,000 in an effort to locate any toxic or hazardous materials. But they have stated that a clean-up plan presented by the DEC is overkill, in that it requires extensive testing throughout the pit.

APPENDIX P

Board values homes near pits at $1

By BERT GRUBB
Staff Writer

The borough's Board of Equalizations decided late Thursday at least three of the residences near Unocal's West Poppy Lane gravel pits were worth $1 apiece because of environmental issues surrounding charges of illegal dumping at the pits.

Borough Assemblyman Dave Carey said the Assembly was acting last week as the Board of Equalization to hear property owners protest their property value assessments. He said the board made the decision on the West Poppy Lane properties after hearing the arguments of three families with residences and land near the pit area.

The state Department of Environmental Conservation presented Unocal with a clean-up plan earlier this month. DEC and Unocal are supposed to meet on the plan this week.

Carey said one of the residents presented documentation of the dumping of various potentially toxic materials at the pit site.

One resident also raised the possibility of radioactive material being among alleged contaminants spread toward residences "every time the wind blows," according to Carey. Carey said one of the residents also displayed a letter in which a local realtor reportedly said the property in the area of the pits was virtually worthless.

Carey said the residents' presentations raised a number of questions as far as he is concerned.

"I would like to know what the borough's responsibility is, particularly if we are granting permits or are involved in permitting for oil exploration," Carey said. "On the local level, we're here to respond to the needs of the people. What is it that the borough should be doing? Have we complained about pollution problems?"

Sterling Assemblyman Fran Moore made a motion to reduce the three property owners' assessments in West Poppy Lane to 10 percent of the value it was given by the borough assessor when the residents came before the BOE, Carey recalled. He said by that time, the assessor had already reduced the property 50 percent below its 1986 assessed valuation.

Carey amended the motion to parcel of land and $1 per residence, and the amended motion passed.

Carey said BOE members discussed the possibility of similarly reducing assessments on all properties within a quarter mile radius of the West Poppy Lane pits. But he said the BOE can only legally reduce assessments on properties whose owners formally appeal to the board. He said only those three, Mr. and Mrs. Ronald Engleman, Joseph and Billie Hardy and Doris Tottin, appealed.

He said the BOE also considered passing a resolution recommending Borough Assessor Don within a quarter of a mile of the gravel pits, but felt it didn't have the authority. He said such a resolution might be something the Assembly should consider.

Carey said he has asked the borough attorney's office to see whether the borough has grounds for a lawsuit against Unocal, since the borough owns lands in the vicinity of the pits. He feels the value of the borough lands may also have been damaged.

"I would like to hear more from the borough about the danger to residents and the borough's liability," Carey said. "I don't know what we should do. There are pits

Ronald Engleman, one of the successful West Poppy Lane protestors whose property valuation was reduced, said he's lived in the West Poppy Lane area for quite awhile.

"If I had the money, I'd be out of here tomorrow," Engleman said. "But everything I own is tied up here. I can't give it away. And what's scary is nobody will call it a problem."

He hopes the BOE's action will call attention to the situation.

"I'm just glad somebody is finally realizing there is a problem here," he said.

It's time for action on Kenai gravel pit

The state of Alaska is ready to take Union Oil of California to court over a Kenai Peninsula gravel pit—and rightly so. Department of Environmental Conservation officials say the pit, near Kenai Peninsula Community College, has been contaminated by Union. These officials say their decision to sue is final unless Union quickly shows them a cleanup plan to remove the contaminant, a byproduct of natural gas production similar to kerosene.

Union disagrees. A company spokesman says it's time for more talk between the state and the company and more tests because some important questions remain unanswered. In Union's judgment, the state's tests either exonerate the company—or demonstrate nobody has the answer yet.

But the record suggests it's time for action, not talk. Local people, concerned about the impact of chemicals from the pit, fear their drinking water could be contaminated. They should not have to endure such fear, although Union and state tests now indicate nearby wells are safe. Moreover, dangerous drilling mud has been discovered in one section of the pit—and DEC officials also found globules of an unknown industrial waste oozing from the ground.

Given these circumstances, Union's call for study and delay clearly is unreasonable. This is no time for more study.

Alaskans have been led to believe that industrial pollution happens only Outside, in backward states where people don't care about their environment or have become habituated to oily water and dirty air. Yet the reality is that pollution can happen anywhere when the public is unobservant and undemanding.

Union has had plenty of opportunity to address the problems created by these pollutants. They haven't done so. It's time for the state to step in and require them to clean up this piece of the Kenai Peninsula without delay.

APPENDIX Q

STATE OF ALASKA

JAY S. HAMMOND,
GOVERNOR

DEPT. OF ENVIRONMENTAL CONSERVATION
KENAI FIELD OFFICE

P.O. BOX 1207
SOLDOTNA, ALASKA 99669

April 5, 1982

Mr. John Woodruff
c/o Seward Fisheries
P.O. Box 7
Seward, AK 99664

Dear Sir:

It appears that Alaska Environmental Industries will soon have the capabilities of refining Tesoro's residual sludge to a state where it can be utilized in your boiler burners. In a recent demonstration conducted in my presence of burning this material in your boiler burners, it certainly appeared that this indeed is possible. It was noted that no black smoke was present in this trial burning. The A.D.E.C. has always encouraged projects such as this where a waste material can be converted into a useful commodity. In my discussions with Tesoro's environmental engineers and chemists, I am told that in burning this waste there will not be anymore of a problem with air contaminants than what is being burned at the present time (crude).

The A.D.E.C. has no objection in burning this material in boiler burners. We will expect that the air quality be maintained. We may also want to conduct some air contaminants emission tests in the future, as this is the first time that this material has been utilized as a fuel. If this program is as successful as it appears, I can foresee your firm converting your other facilities to this type of fuel.

Sincerely,

Howard R. Keiser
Environmental Field Officer

cc: Tesoro
Alaska Envir. Ind., Inc.

HRK:jlg

APPENDIX Q

STATE OF ALASKA

JAY S. HAMMOND,
GOVERNOR

**DEPT. OF ENVIRONMENTAL CONSERVATION
SOUTHCENTRAL REGIONAL OFFICE**

437 E. STREET
SECOND FLOOR
ANCHORAGE, ALASKA 99501
(907) 274-2533

P.O. BOX 615
KODIAK, ALASKA 99015
(907) 486-3350

P.O. BOX 1207
SOLDOTNA, ALASKA 99669
(907) 262-5210

P.O. BOX 1709
VALDEZ, ALASKA 99686
(907) 835-4698

P.O. BOX 1064
WASILLA, ALASKA 99687
(907) 376-5038

April 5, 1982

Mr. Glen Jackson
Tesoro Alaska Refinery
P.O. Box 3691
Kenai, Alaska 99611

Dear Mr. Jackson;

I have been working with Alaska Environmental Industries Inc. formerly Mar Enterprises Inc. for sometime exploring the possibility of utilizing Tesoro residual sludge as a fuel in Seward Fisheries, boilers in Seward, Alaska.

Two different demonstrations reveals that this material can be utilized as a boiler fuel. In the demonstrations no more black smoke was noted than when the boilers were using regular fuel.

This department has always encouraged programs such as this where a waste can be converted to a useful commodity.

APPENDIX Q

A.E.I. will assume all responsibility for handling this waste from the time they pick up load until its ultimate destination and usage.

D.E.C. will hold A.E.I. to this policy. A.E.I. will of course be expected to abide by current oil spill reporting and clean up procedures.

If this office can be of further assistance please call 262-5210.

Sincerely,

Howard R. Keiser
Environmental Field Officer

cc: A.E.I. Inc.
Seward Fisheries Inc.
HRK:jig

APPENDIX R

STATE OF ALASKA

JAY S. HAMMOND,
GOVERNOR

DEPT. OF ENVIRONMENTAL CONSERVATION
KENAI FIELD OFFICE
Tel.:

P.O. BOX 1207
SOLDOTNA, ALASKA 99669
(907) *262-5210*

December 1, 1980

MAR Enterprises, Inc.
P.O. Box 1660
Soldotna, AK 99669

Re: Open Burning of Drag Reducing Agent from Alyeska

Dear Sirs:

Enclosed please find a copy of: (18 AAC 50.030. Open Burning, and 18 AAC 50.110. Air Pollution Prohibited,) which are concerned with your request for our review of this burning.

Open burning is permissible. In this instance the following conditions shall be adhered to:

1) Maximum combustion efficiency shall be maintained throughout the burning period (using auxiliary air flow).
2) Reasonable procedures shall be used to minimize the amount of smoke generated, and no black smoke shall be allowable.

    a) minimize spillage and associated black smoke,

    b) use of adequate extinguishing methods.

APPENDIX R

3) Storage and transportation in covered containers is necessary to prevent spillage.

Sincerely,

Roger Moulton
Environmental Engineer

rms
Enclosure

SEE PERMIT
PAGE 2
PARAGRAPH 2

APPENDIX R

PERMIT CONDITIONS AND STIPULATIONS

A. OPERATION—The permittee shall:

1. Accept only the following materials for processing and disposal:

   a. Waters containing less than four percent oil except emulsion may exceed four percent.

   b. Solid waste contaminated with petroleum and its products.

   c. Non-toxic water conditioning sludges.

2. Be allowed to accept other wastes only after prior written permission has been obtained from the department's Southcentral Regional Office in Anchorage or the Central Office in Juneau. Retroactive permission will not be granted.

3. Place oil contaminated solids in Trench No. 1, in lifts not to exceed six (6) inches in depth. Each lift shall be thoroughly mixed with native soils to a consistency capable of supporting heavy equipment traffic before the next lift is placed.

4. Place no material in the winter storage areas or the watery wastes trench at an elevation higher than the elevation of the original ground surface unless the engineering design of the surrounding dikes has been approved by the department.

5. Reserve at least one watery wastes pit or similar, approved structure each fall for the deposition of these wastes each winter.

6. Be allowed to accept waste oils for temporary storage only. Storage shall be in a steel tank or other approved storage structure.

7. Operate and maintain the facility in accordance with the attached portions of the permit application except as noted herein.

B. Monitoring and Reporting—The permittee shall:

1. Install groundwater monitoring wells in accordance with the recommendations of the EPA Region X technical assistance team scheduled to examine the facility during the spring of

1979. Installation shall be completed within sixty days of receipt of EPA's recommendation.

2. Monitor the groundwater parameters beneath the site as recommended by the EPA technical assistance team.
3. Maintain a log of all wastes received at the facility. The log shall include waste type, quantity, source, date and disposal method. A copy of the log shall be submitted each January to the department's Southcentral Office in Anchorage.

APPENDIX S

STATE OF ALASKA

BILL SHEFFIELD,
GOVERNOR

DEPT. OF ENVIRONMENTAL CONSERVATION
SOUTHCENTRAL REGIONAL OFFICE

437 E. STREET
SECOND FLOOR
ANCHORAGE, ALASKA 99501
(907) 274-2533

P.O. BOX 615
KODIAK, ALASKA 99615
(907) 486-3350

P.O. BOX 1207
SOLDOTNA, ALASKA 99669
(907) 262-5210

P.O. BOX 1709
VALDEZ, ALASKA 99686
(907) 835-4698

P.O. BOX 1064
WASILLA, ALASKA 99687
(907) 376-5038

February 28, 1983

Mr. Dave Brown
MAR Enterprises
P.O. Box 1660
Soldotna, Alaska

Mr. Brown:

The evaluation of the test burn of the drag reducing fluid on February 14, 1983, demonstrated that, currently, you do not have the ability to burn this material without creating black smoke. Written approval is therefore required under 18 AAC 50.030(b) before any additional material may be burned. Approval to open burn material that gives off black smoke is based on a review of data submitted by the applicant, historical data available to the Department, best management practices and health and environmental effects on the local population. Minimum data from the applicant includes type and amounts of material to be burned, methods used to burn the material, method employed to reduce black smoke and how the material would be extinguished to stop an unexpected threat to health or environment. Second, the storage/disposal site is a solid waste disposal facility as defined in 18 AAC 60.130(16) and must have a permit to continue operation. I have also enclosed a Solid Waste Management Permit Application for your use. The application explains what information needs to be submitted in order for a review to take place.

Sincerely,

Robert C. Flint
Regional Program Coordinator

ENCLOSED: 18 AAC.50 Air Quality
18 AAC.60 Solid Waste Management
Solid Waste Management
Permit Application

cc: Kenai District Office, Bob Cannone
Carl H. Harmon, ADEC, Anchorage
Bill MacClarence, ADEC, Anchorage

APPENDIX S

STATE OF ALASKA

BILL SHEFFIELD,
GOVERNOR

DEPT. OF ENVIRONMENTAL CONSERVATION

Telephone: (907) 262-5210
Address: P O Box 1207
Soldotna, AK
99669

May 6, 1986

Dave Brown, President
Alaska Environmental Industries
P.O. Box 1660
Soldotna, AK 99669

Re: Compliance Order by Consent 86-01

Dear Mr. Brown:

Your plans for the disposal of Drag Reducing Agent spilled onto the ground surface at Arness's property, for the disposal of contaminated soil at Arness's and Mansfield's property and buried D.R.A. at Mansfield's property have been reviewed by the Department.

The Department cannot approve the disposal of this material at the Sterling Special Waste Site since it was not permitted for this purpose. Disposal of this material at the Sterling Special Waste Site would be a violation of the permit. We do approve, however, of this material being shipped to a proper waste disposal facility as you have indicated in your letter dated 12/27/85 and 4/15/86. Since there is no disposal facility located in the state of Alaska for this waste material, it may be necessary to ship it out-of-state. Please advise this office which disposal facility you intend to use and the date by which this will be accomplished.

Sincerely,

____

Environmental Field Officer

ML/me

cc: Bill Lamoreaux ADEC, SCRO
Bruce Erickson ADEC, SCRO
Michelle Brown
Attorney General's Office
1031 W. 4th Ave. Suite 200
Anchorage, AK 99503

APPENDIX T

STATE OF ALASKA

BILL SHEFFIELD, GOVERNOR

DEPT. OF ENVIRONMENTAL CONSERVATION

Telephone: (907) 262 5210
Address: P O Box 1207
Soldotna, AK
99669

November 8, 1985

Mr. Ben Hilliker
Alyeska Pipeline Service Company
1835 S. Bragaw
Anchorage, Alaska 99512

Re: Drag Reducing Agent

Dear Mr. Hilliker:

The Department would like information from you indicating the amount of money paid Dave Brown of Alaska Environmental Industries, and formerly M.A.R. Enterprises, for the disposal of your Drag Reducing Agent. By this letter we are requesting that you provide us with copies of all invoices, receipts, etc. that indicate these dollar amounts.

Thank you for your assistance.

Sincerely,

Environmental Field Officer

me

APPENDIX U

July 28, 1986

To Whom it may concern:

I, Walter E. Ellis, am a registered Mechanical Professional Engineer, registry number 14805, State of Ohio.

I have personally inspected the waste water pit in the north west corner of the Sterling Special Water Site at Sterling, Alaska, on July 28, 1986.

The pond bottom appeared level and rectangular measuring 95 feet 6 inches by 61 feet 6 inches. This level pit bottom area was confirmed by the existing oil stains.

My calculations, attached, indicates that a discharge of 60 barrels or 2520 gallons would raise the level of the pond 0.6436 inches (Sixty Four Hundred Thirty Six Ten Thousandths of one inch)

To get an oily slick or sheen 2″ deep would require 7831 (Seven Thousand Eight Hundred and Thirty One) gallon.

To get an oily slick or sheen 3″ deep would require 11,746 (Eleven Thousand Seven Hundred and Forty Six) gallon.

Walter E. Ellis, P.E.

Witnessed by: Constance L. Hobbs
Postmaster Girdwood AK

Addendum to Letter of July 28, 1986

Calculations:

Area = 95′-6″ × 61′-6″ = 5873.25 sq. ft.
5873.25 × 144 sq in/sq ft = 845/48 sq in.

2520 gal 8.0 gal/cu ft

$$\frac{2520\text{ gal}}{8.0\text{ gal/cu ft}} \times 1728\text{ cu in/cu ft} = 544320\text{ cu in.}$$

$$\frac{544320\text{ cu in}}{845748\text{ sq in}} = 0.6436\text{ inches deep.}$$

APPENDIX U

If there were 2″ of liquid over the area observed, then:

Area = 845748 sq in × 2 inches deep = 169,486 cu in.

$$1 \text{ gal} = \frac{1728 \text{ cu in/cu ft.}}{8.0 \text{ gal/cu ft}} = 216 \text{ cu in/gal.}$$

$$\frac{1691496 \text{ cu in.}}{216 \text{ cu in/gal}} = 7831 \text{ gallon for 2″ deep.}$$

If there were 3″ of liquid over the area observed, then:

Area = 845748 sq in × 3″ deep = 2537244 cu in.

1 gal = 216 cu in/gal

$$\frac{2537244 \text{ cu in}}{216 \text{ cu in/gal}} = 11746 \text{ gallon.}$$

Witnessed by: Constance L. Hibbs

Walter E. Ellis, P.E.
Postmaster
Girdwood, AK 99587

APPENDIX V

CONFIDENTIAL
ATTORNEY/CLIENT
COMMUNICATION

STATEMENT BY ANTHONY ZENK
8/21/85

I ANTHONY ZENK received one call from Ann Toppa that said there will be a truck at the Sterling Special Waste Site at 10:00 am and to tell Mike that the "*Deal happened.*" I informed this person that Mike would return at 1:00 pm. This call was received at about 9:30 am.

I received a call from Dan Roberts at about 9:45 am, as I was answering the phone, and I don't recall who he asked for at first, but he did say that an oil truck would be at the Sterling Special waste site at 10:00 am and that someone from DEC should be there to take a sample. He then asked to speak to Ted and said that it was very important. I put him on hold, described to Ted what was said, and returned to the phone and told Dan Roberts or person calling, that if he would hold Ted would speak with him. He held for awhile, about 5 minutes and then called back on a different line, then described that he had been cut off, but the light was still blinking. He mentioned that he was still interested in speaking with Ted, I put him on hold again. In about 2 minutes as he was holding, I then intercepted and told the caller "Dan Roberts," that Ted was busy and would call him back if he did not want to hold. Caller, "Dan Roberts" then said someone should investigate and hung up.

ANTHONY ZENK

APPENDIX W

**CONFIDENTIAL
ATTORNEY/CLIENT
COMMUNICATION**

STATEMENT BY DAN ROBERTS
8/23/85

C: August 23, 1985, 10.15 present are________, ________________. Statement from Dan Roberts on the events that occurred August 21, 1985

Roberts: I'm Dan Roberts and on this particular date I noted by time each event. But let me back up to the day before, which had been 8/20/85, I received a phone call from, Construction Services Inc. had received a phone call from an oil company requesting Construction Services to come out and pump the bilge on one of their barges. I told the oil representative that we couldn't pump it any more, that we had oil on hand that we couldn't get rid of and that we had to stop pumping the bilges because . . .

B: Do you know what oil company that was and what representative that was?

Roberts: I do, I had, I don't think that is pertinent to this thing, I'd just as soon leave that out if it's all right with you guys.

C: Sure

Roberts: At least lets leave it out for now and if you come along later and think it's necessary, then we'll . . .

Ah, anyway, I made the statement to him that he might call Dave Brown of Alaska Industries or whatever they are, I notice they've had a name change in the paper. Anyway, I talked back with this oil company representative later and asked him a question, did Alaska Environmental Industries say they could take this, whatever was in the bilge? And he said yeah, they said they could, no problem. Well, we pumped the bilge, not many times, but a few times, and it's always got a higher percent than the four percent oil, so it couldn't go to the Sterling Waste Site. However, it doesn't usually have an awful lot of oil in it, anywhere from 30 gallons to, it could be as high as 6 or 8 barrels total volume and usually somewhere 30, 40 barrels.

B: Of bilge or oil?

Roberts: This is total. What I was saying before was the amount of oil was in there and the way we came up with that was we put this in a tank which I understand is the correct procedure to take, we put this oil into a tank and let the water fall out of it and then measure the oil and measure the water how much it is. I tell you that happened on the 20th, and in trying out that they weren't going to, that AEI was going to haul this bilge, I decided that due to things that's happened in the past between Alaska Environmental Industries, Dave Brown, that he has pulled on us through articles in the paper and such

APPENDIX W

as, it was time to see if we couldn't do something back. A problem here if there was one existing, and that problem I'm speaking of, if there was too much oil . . .

So the next morning I still has this in mind, and with a camera I went out to the dock. I viewed AEI's truck with a hose going down to the barge there at the Nikiski dock.

APPENDIX AAA

MAR ENTERPRISES, INC. Alaska 343
BOX 1034 Statebank
ALASKA Soldotna, 99669

Pay Four Thousand Nine Hundred Fifty-Five Dollars

		DATE	AMOUNT
TO THE ORDER OF	Bar	1/9/78	$4,955.00

NOT NEGOTIABLE

This bar was not in the real estate business, nor was Dan Lyan.

The people on the Kenai Peninsula know who Dan Lyan is and they know which bar this is.

CHECK NO	DATE	CHECK ISSUED TO	IN PAYMENT OF	AMOUNT OF CHECK	balance
343	1/9/78	Bar (commission)		4,955.00	20,673.03

APPENDIX BBB

Fact Sheet

United States Environmental Protection Agency
Region 10
Park Place Building, 13th Floor
1200 Sixth Avenue, WD-137
Seattle, Washington 98101
(206) 442-8155

Public Notice Issuance Date: July 26, 1988
Public Notice Expiration Date: August 25, 1988

Permit No.: AK-000050-7

PROPOSED REISSUANCE OF A NATIONAL POLLUTANT DISCHARGE ELIMINATION SYSTEM (NPDES) PERMIT TO DISCHARGE POLLUTANTS PURSUANT TO THE PROVISIONS OF THE CLEAN WATER ACT

Unocal Chemicals Division—Kenai Plant
Post Office Box 575
Kenai, Alaska 99611

has applied for reissuance of a National Pollutant Discharge Elimination System (NPDES) permit to discharge pollutants pursuant to the provisions of the Clean Water Act. This fact sheet includes (a) the tentative determination of the Environmental Protection Agency (EPA) to reissue the permit, (b) information on public comment, public hearing and appeal procedures, (c) the description of the current discharge, (d) a listing of tentative effluent limitations, schedules of compliance and other conditions, and (e) a sketch or detailed description of the discharge location. We call your special attention to the technical material presented in the latter part of this document.

Persons wishing to comment on the tentative determinations contained in the proposed permit reissuance may do so by the expiration date of the Public Notice. All written comments should be submitted to EPA as described in the Public Comments Section of the attached Public Notice.

After the expiration date of the Public Notice, the Director, Water Division, will make final determinations with respect to the permit reissuance. The tentative determinations contained in the draft permit will become final conditions if no substantive comments are received during the Public Notice period.

APPENDIX BBB

The permit will become effective 30 days after the final determinations are made, unless a request for an evidentiary hearing is submitted within 30 days after receipt of the final determinations.

The proposed NPDES permit and other related documents are on file and may be inspected at the above address any time between 8:30 a.m. and 4:00 p.m., Monday through Friday. Copies and other information may be requested by writing to EPA at the above address to the attention of the Ocean Programs . . .

Oily waste skimmed from the effluent basins is no longer sent to the Sterling disposal site, which was closed in 1985. Unocal now burns all waste oil in one of its boilers under a state air quality permit.

E. *Compliance History*

The Unocal plant has an excellent history of compliance. The previous permit contained effluent limitations on pH, oil and grease, organic nitrogen, and ammonia nitrogen. *No permit violations occurred during the five-year period covered by the previous permit.* (Two reported exceedances of the pH range of 6.0–9.0, for December 1983 and August 1985, did not last long enough to qualify as violations.) . . .

The dewatering system shown in the application is not yet fully functional. *The Kenai Borough landfill has refused to accept Unocal's sludge in the past because of its high water content.* The company found that centrifugation and belt processing were not effective in removing the water. They recently found that freeze-drying is promising. To date, they have pumped the sludge into gravel pits at the end of the summer, then allowed the liquid to evaporate, or freeze and sublime in cold weather. Remaining liquid drains into a sump, which is pumped out. The residual sludge is granular and nontoxic. *The company will again ask the Borough about sludge disposal this year.*

APPENDIX CCC

ALASKA ENVIRONMENTAL INDUSTRIES, INC.

Industrial Vacuum Pumping
Waste Disposal
Tank Cleaning
Tank Liners
Oil & Chemical Spill Cleanup

Formerly Mar Enterprises
Box 1660 • Soldotna, Alaska 99669

Bus: 283-5050 or 283-4906
Res. 262-2863

Power Washing
Roustabouts
Power Washer &
Steam Cleaner Sales
Service & Rentals

August 5, 1986

Office of the Governor
3rd Floor State Capitol Building
Box A
Juneau, Alaska 99811

Dear Governor Sheffield,

No one has ever offered a satisfactory explanation for thirteen years of deliberate nonfeasance by D.E.C. where the oil industry is concerned. We are looking at unlawful disposal, no permits, no liners, no enforcement, no penalties. The situation smells more of possible corruption than of being overworked and under staffed.

Over the years, I have heard rumors of oil industry payoffs to D.E.C. employees, but I have never known whether they were true or false.

In the last few months I have been told this same story on five separate occasions, by three different individuals. I have no reason to doubt the integrity of any of these people.

I have been told what oil company it is that is allegedly making the payoffs. I have been told the name of the employee of that oil company who allegedly acts as the "bag man" and makes those alleged payoffs. One of my sources says he has actual proof, not hearsay.

I don't know whether the allegations are true or not. I only know that they are serious enough to warrant an immediate investigation by the office of the Attorney General, not by the D.E.C. D.E.C. covers up their own investigations "in house."

When D.E.C. employees regularly use state vehicles seven days a week for personal use in direct violation of Chapter 68, Section 44.68.010, it is grounds for dismissal of the offending employees, but it goes on day after day and Ross & Martin cover it up. I have personally complained to Bill Ross and Keith Kelton about the personal use of state vehicles and state credit cards, but no action was ever taken. That is stealing from the taxpayers, Governor Sheffield.

When a top ranking D.E.C. employee stores furniture belonging to his girlfriend in a warehouse in Anchorage, under the guise of D.E.C. tools and equipment, and the D.E.C. pays the storage rent for a year or more, that, Governor Sheffield, is also stealing from the taxpayers.

When that same top ranking D.E.C. employee gives that same girlfriend a large undeserved pay raise then I would say that that particular D.E.C. employee had better start paying for his own love life, instead of expecting the taxpayers to pick up the tab.

When that same top ranking D.E.C. employee lies three times in a memo on August 28, 1984, with copies of that memo going to two of our Governor's top aides, then as far as I am concerned, that man lied to you, Governor Sheffield.

When ______________ was made aware of the aforementioned situations, together with ______________, by me personally, and when they did nothing about it, they then are guilty of covering up for their friends.

When in the aforementioned memo, the lies refer to the disposal practices of certain segments of the oil industry then why lie, unless perhaps there is some truth to the allegations of payoffs.

In 1982, one D.E.C. employee was in a bit of hot water and was somewhat overdrawn at his bank. Within

APPENDIX CCC

ninety days after cleaning up the mess, he bought a brand new 3/4 ton pickup and a new 10 foot camper. He bought a new 23 foot boat, motor and trailer during the same 90 days. He also bought a building lot during this 90 days. None of this was unique and there may be justifiable reasons, but after the "hot checks" , the sudden wealth was considered strange on the Peninsula. One year later that D.E.C. employee built a new home on that lot with a $123,000 deed of trust. The same D.E.C. employee also bought land and a cabin in Homer at about the same time.

I have two computer printouts on this particular D.E.C. employee, one on his real estate interests and one from the Department of Motor Vehicles. The real estate printout is nine feet long from 1982 to today, and it indicates he owns property in Homer, Seward and Kenai/Soldotna. It also lists his deeds of trust.

The D.M.V. printout indicates he is making payments on two vehicles, plus a boat, motor and trailer. Of course he is also paying on his real estate.

As near as I can estimate, this young man and his wife have been paying somewhere in the neighborhood of $2500 to $3000 per month for toys for the last four years, not bad considering it was instant, overnight wealth.

I know the young man's wife works too, but I also know what taxes and other deductions take out of pay checks. I also know they take annual vacations to Mexico and Hawaii.

I am not accusing this young man of anything, but like everyone else who knows him, I am wondering how he lives like a millionaire on a D.E.C. field officer's salary, when four years ago he couldn't manage his money.

I am also wondering why this individual should now find it necessary to sell his home and resign from his job. I wonder why he is resigning and his wife is resigning and why they are leaving when Alaska has been so good to them.

APPENDIX CCC

One high ranking D.E.C. employee lied under oath, four times in Anchorage at the hearings concerning the permit for the Sterling Special Waste Site. That is perjury, Governor Sheffield, and perjury is a felony. I wonder if that individual realizes that convicted felons lose the right to vote, lose the right to possess firearms, lose the right to do a lot of things, usually lose their job too.

Governor Sheffield, in the Alaska Department of Environmental Conservation today, there is very little integrity or credibility.

I think it's time that you and/or Attorney General Brown contacted me. We have a lot to talk about.

Thank you kindly.

Sincerely,

Dave Brown

APPENDIX CCC

DAVID A. BROWN
Alaska Environmental Industries Inc.
Box 1660
Soldotna, Alaska 99669

February 4, 1987

Governor Steve Cowper
State of Alaska
Office of the Governor
P.O. Box A
Juneau, Alaska 99801-0101

Dear Governor Cowper:

First of all let me congratulate you on running a clean campaign and on winning the race. So far I have been very favorably impressed by your honesty and your decisiveness.

I am sure, Governor Cowper, that you remember meeting me in Soldotna outside the Sports Arena before the primary election and again in your office in Anchorage in the old Club 25.

I was favorably impressed with your recent handling of the Mike Bradner situation. My question is this. If your administration feels it cannot afford even the appearance of impropriety, then why is   still employed by the DEC when April Cook, the Investigator for the AG's office, has already established his lack of integrity in the matter of the illegal storage of his girl friend's furniture?

Why is _______________ still employed by the DEC when I have charged him with three (3) counts of perjury at my June permit hearing? Why are _______________, _______________ and _______________ all still employed by the DEC when all three of them are guilty of regular, unlawful personal use of state-owned vehicles?

Why is _______________ still employed by DEC against his own wishes? _______________ sold his home and other assets, resigned, and was leaving; but, was persuaded by DEC (apparently _______________) to stay. ______________ seems to fear something, like perhaps if he leaves he might be made a scapegoat for all of the DEC's transgressions and non-feasance. If the state is short of money and ______________ wants to go, he should be allowed to go. I think there is a lot that ______________ could tell if he were approached right and offered immunity.

No one seems to want to know the truth, for the truth might point to people in high places in past administrations.

The truth is that while DEC condoned unlawful disposal on the Kenai Peninsula, the oil industry has saved approximately $20,000,000 and probably $200,000,000 statewide. I did not get any of that money, even though I operated the only permitted drilling mud dump site in the state. Who got wined and dined for condoning all that illegal disposal? And all the the trips and tips, who got them? Not me.

Why did IRS demand and get from me recently all of my computer printouts on ______________? Why should IRS feel that ______________ life style and his tax returns don't match?

I believe that the closure of the Sterling Special Waste Site was bought and paid for by the oil industry with the help of the DEC, and possibly others unknown in Juneau.

Why is it that pollution is a crime in every state except Alaska? Alaska rewards polluters.

Why has unlawful pollution been accepted and condoned even encouraged by ______________, ______________, ______________, ______________ and ______________? For six and a half years I complained to DEC about unlawful dumping and pollution by the oil industry. Nothing was ever done. Why?

Any complaint I made to DEC regarding unlawful disposal by the oil industry only resulted in uncalled for harassment, of me, by DEC, with nothing done by DEC to the offending oil companies.

This led me to believe long ago, that some employees of DEC might have a second income from the oil industry. Now I have considerable evidence to support that belief.

Additionally, it is no secret that our former governor and certain legislators accepted oil industry favors, if in no other form than free airplane rides around the country and a few fundraising parties. That is pretty stiff competition for a small businessman.

You were not in Juneau when all of this took place, so I know your record is clean. Therefore, since some honest answers and a full investigation cannot possibly harm you, I think it is time we, or I, got those answers. There is far more to this problem than mere incompetence or lack of manpower and funds. The problem starts with lack of integrity in the aforementioned individuals.

I have enclosed for your review, so you may gain some insight into this problem, two letters to former Governor Sheffield, a letter to the Ombudsman and a memo from ____________ to ____________, referred to by me in my letter to then Governor Sheffield on October 22, 1986.

I shall be anxiously awaiting your reply, Governor Cowper, and I shall be happy to cooperate in any investigation which you may choose to initiate.

Thank you kindly for your consideration.

Sincerely,

David Brown

DAB:ds

APPENDIX CCC

Box 1660
Soldotna, Alaska 99669
February 28, 1987

Governor Steve Cowper
P.O. Box A
Juneau, Alaska 99811

Dear Governor Cowper:

Earlier this month, February 3 or 4, I wrote to you asking why certain people were still employed by the D.E.C., in as much as their lack of integrity was already a fact proven to you.

As yet I have received no answer from you, but worse than that, I no longer have my copy of that letter. I dislike not having a record of what I said. I would appreciate it if you would kindly send me a xerox copy of my letter to you.

I didn't lose the letter. Thinking I had a copy, I personally hand carried the letter to United States Congressman Morris K. Udall from Arizona, at his office in Phoenix, Arizona. I also delivered to Congressman Udall, a copy of my lawsuit and a rather thick package on the unlawful disposal of drilling mud by the Oil Industry in Alaska and the failure of the State of Alaska to enforce the Statutes and Regulations that have existed since 1973, regarding said disposal.

I felt that in as much as the Oil Industry and the State of Alaska have exhibited no concern for the environment, or for the people of the State of Alaska or for future generations, then perhaps it was time to take other steps.

By now I am quite sure that in Washington, D.C., the entire congressional subcommittee which has the power to issue or deny the Arctic National Wildlife Refuge permit, has been well informed as to the mess that past administrations in the State of Alaska have allowed the Oil Industry to make of our beautiful State, all in the interest of their combined greed.

I would imagine by now that the powerful environmental lobby in Washington, D.C. may possibly have the facts also, as the facts actually exist, rather than the propaganda which they have been fed in the past. If not, then probably they soon will.

Governor Cowper, please send me a copy of my letter, and if time allows, then please also send me an answer to that same letter.

Thank you kindly.

Sincerely,

Dave Brown
Alaska Environmental Industries Inc.

cc: Congressman Morris K. Udall
Senator Ted Stevens
Senator Frank Murkowski
Congressman Don Young

APPENDIX CCC

Box 1660
Soldotna, Alaska 99669
October 22, 1986

Office of the Governor
3rd Floor
State Capitol Building, Box A
Juneau, Alaska 99811

Dear Governor Sheffield:

Regarding my letter to you, dated August 5, 1986 and the subsequent investigation by April Cook initiated at your request, I have some concerns.

First of all, let me state that I am grateful to you for initiating the investigation. My concerns are the following:

By now, April Cook must know whether or not ______________ did indeed store his girlfriend's furniture in an Anchorage warehouse under the guise of it being D.E.C. tools and equipment, with the D.E.C. paying the storage rent for a year or more. Perhaps Mr. ______________ is not guilty, but I have been told this story by two different D.E.C. employees and I have no reason to doubt their word. If the incident did indeed occur, then it does not matter that the amount was small. Whether it was $50 a month, $75, or $100 a month, it still amounts, I believe, to a felony, usually described as Grand Larceny. Perhaps I am wrong. Am I wrong? If the crime never occurred, then I owe Mr. ______________ an apology. If the crime did occur, then Mr. ______________ should be fired with no possibility of rehire ever by the State of Alaska.

In my letter to you of August 5, 1986 I refer to Mr. ______________ lying three times in a memo dated August 28, 1984. I have enclosed for your review, Governor Sheffield, a copy of that memo with my comments relating to the memo:

First Lie Paragraph 1, page 1, "We have repeatedly told Mr. Brown we will investigate any incident of which he is aware."

The Truth I, Dave Brown, have personally complained to D.E.C. approximately 100 times about il-

APPENDIX CCC

legal mud disposal. I complained to Martin, Lamoreaux, Flint, Cannone, Lucky, Keiser, Moulton, and Harmon. Not one single investigation was ever initiated. ______________ is a liar.

Finally in January 1985 after becoming fed up with D.E.C. lies and incompetence, I dug up Cannery 2, Union Oil's mud pit on Mike Pelch's land on Beaver Loop Road.

After exposing 34,000 barrels of illegally buried drilling mud, I called in four newspapers, one T.V. station, the E.P.A. and the Corps of Engineers, as well as the D.E.C. That fiasco cost me, Governor Sheffield, over $4,000 out of my own pocket, to expose not only Union Oil, but also the D.E.C. for their total nonfeasance. Then, and only then, did the State of Alaska decide to file suit against Union Oil, but not because the State wanted to sue, only because the State was embarrassed into action.

Second ______________ *Lie* Page 1, item 3 under paragraph 2, "The reserve pits used for disposal are fully lined."

The Truth Governor Sheffield, out of over 300 mud pits in the Kenai Peninsula Borough, only two pits are fully lined, Chevron at Beluga and ARCO at Wolf Lake 2. 298 or more pits are unlined.

Third ______________ *Lie* Also page 1, item 3 under paragraph 2, "The reserve pits are sealed with an impermeable cap."

The Truth Governor Sheffield, out of the more than 300 mud pits previously mentioned, two pits are sealed with impermeable caps. Both of the capped pits belong to ARCO, one at Funny River and one at Wolf Lake 2. All of the remaining 300 or so pits are backfilled with tree trunks stumps, trash, and dirt. Some pits although abandoned are not capped or backfilled.

As to the rest of Mr. ______________ memo, I have a few more comments.

Item 2 "Under the control of the entity producing the waste" Governor Sheffield, if you can't trust the entity producing the waste to get a disposal permit or to even apply for a permit, then that entity cannot be trusted to dispose of their own waste.

Item 5 "Disposal is at the site of waste production and thus eliminates the need for additional handling and transport for disposal elsewhere. Governor Sheffield, septic waste is not disposed of at the site of waste production. Garbage is not disposed of at the site of production. Both are handled and transported for disposal elsewhere.

Drilling mud, in every other state except Alaska, is handled additionally and transported for disposal elsewhere. In California, it is common to haul drilling mud as far as fifty (50) miles one way for final disposal, one hundred miles round trip. Also in California, occasionally drilling mud is hauled one hundred miles (100) one way, two hundred miles round trip.

Just who is ______________ working for anyway the people of the State of Alaska, or the oil industry?

Bill Ross may hold the title of D.E.C. Commissioner just as did Richard Neve and Ernst Mueller, but ______________ runs the D.E.C. and he always has. All that the D.E.C. Commissioner in Juneau knows is what ______________ ______________ chooses to tell him, and ______________ doesn't always choose to tell the truth.

Copies of this memo, Governor Sheffield, went to Ray Gillespie and Lennie Boston, two of your trusted aides. Gillespie and Boston could not present a true picture to you, because they were lied to by ______________.

Copies of the memo went to ______________ and ______________. They both knew the memo was not true, nor even intelligent, but they did nothing to prevent you, Governor Sheffield, from being further embarrassed by more D.E.C. nonfeasance.

And, Governor Sheffield, your trusted Commissioner, ______________, covered up for his old buddy, ______________, by not bringing these facts to your attention when I pointed out these very same facts to

APPENDIX CCC

the attention of Mr. ______________ in Juneau in the Spring of 1985.

Governor Sheffield, your name was unfortunately linked with the North Slope Borough scandal, the Fairbanks Office Lease scandal, and the high cost of remodeling the Governor's Mansion in Juneau!

When you have high ranking State employees who are incompetent, who lie and cheat, and perhaps steal, and when you then have Commissioners who cover up for these dishonest employees, then it is no wonder that your administration faced problems which it didn't deserve to face.

If there is no substance to the repeated rumors of payoffs to D.E.C. personnel, then why was it necessary for Mr. ______________ to lie so convincingly in the memo of August 28, 1984? Was it just that he wanted to put Dave Brown out of business?

I would hate to think Governor Sheffield, that dishonesty on the part of a State employee would ever be rewarded by continuing job security! As I have said before, there is very little integrity in the D.E.C. The department needs a thorough investigation, plus a house cleaning.

I hope to hear from you soon on this Governor Sheffield. Thank you kindly.

Sincerely,

Dave Brown

APPENDIX DDD

PERMIT #	FILE #	APPLICANT	FACILITY	DATE RECEIVED	DATE DECISION REACHED	DECISION TYPE
Disposal of Heavy Drilling Mud						
8230-BA003	300.15.010	Exxon	Pt. Thompson	8/31/81	12/22/81	Issued
8436-BA009	300.15.031	Chevron U.S.A. Inc.	KIC # 1 Well, Pit 1	9/21/84	12/07/84	Issued
8536-BA004	300.15.026	Alaskan Crude Corporation	ACC Well F-3	2/26/85	4/22/85	Issued
8536-BA005	300.15.042	Alaskan Crude Corporation	ACC Well B-1	3/12/85	11/15/85	Issued
8536-BA006	300.15.045	Alaskan Crude Corporation	ACC Well F-2	2/26/85	4/22/85	Issued
8536-BA007	300.15.043	Alaskan Crude Corporation	ACC Well K-1	3/12/85	11/15/85	Issued
8536-BA017	300.15.031	Chevron U.S.A., Inc.	KIC # 1 Well, Pit 2	9/25/85	11/08/85	Issued
8736-BA011	300.15.097	Texaco, Inc.	Wolfbuttons #1, #2, & #3	11/27/87	6/08/88	Issued
8636-BA020	300.15.064	ARCO Alaska, Inc.	Winter trails #2	11/14/86	1/08/87	Issued
8636-BA021	300.15.065	ARCO Alaska, Inc.	Winter trails #3	11/14/86	1/08/87	Issued
8636-BA022	300.15.066	ARCO Alaska, Inc.	Winter trails #4	11/14/86	1/08/87	Issued
8736-BA001	300.15.072	North Slope Borough	Barrow Gas Field	1/29/87	3/12/87	Issued
8736-BA007	300.15.046	Standard Alaska Productio	W Pad (Eileen West End Pa	7/31/87	9/29/87	Issued
8736-BA008	300.15.047	Standard Alaska Productio	2 Pad (Eileen West End Pa	7/31/87	9/29/87	Issued
8736-BA016	300.15.051	ARCO Alaska, Inc.	Drill Site 3-0	12/18/87	3/22/88	Issued
8636-BA004	300.15.044	Alaskan Crude Corporation	ACC Well F-3	3/13/85	5/05/86	Issued
8636-BA005	300.15.042	Alaskan Crude Corporation	ACC Well G-1	3/13/85	5/05/86	Issued
8636-BA006	300.15.045	Alaskan Crude Corporation	ACC Well F-2	3/13/85	5/05/86	Issued

APPENDIX DDD

8636-BA007	300.15.043	Alaskan Crude Corporation	ACC Well K-1	3/13/85	5/05/86	Issued
9536-BA008	300.15.046	Standard Alaskan Productio	Eileen Well Pad #1	3/26/85	1/22/86	Issued
9536-BA009	300.15.047	Standard Alaskan Productio	Eileen Well Pad #2	3/26/85	1/22/86	Issued

DEPARTMENT OF ENVIRONMENTAL CONSERVATION

Drill Mud Disposal Facilities and Applications

SOUTHCENTRAL REGION

File: WASTE Page: 2

Printed on 6/28/88
Selection Criteria: DRILMUDA Yes
Regional Solid Waste Program Manager:
Henry Friedman (907) 563-6529

Facility Name	Permittee's Name	Permit No.	Issued Date	Expiration Date	Status	Date of Status Determination
Swanson Oil Field	Chevron	8423-BA014			denied	
Trading Bay	Marathon Oil Col	8223-BA010	1/12/83	10/01/1987	temporarily closed	
Wolf Lake #1	ARCO	8223-BA012	11/08/82	10/01/1987	closed	
Wolf Lake #2	ARCO	8423-BA102	5/17/85	5/17/1990	closed	

Total Records Displayed: 20

Disposal of Waste Water by Injection

Facility Name	Permit No.	Location	Issue Date	Expiration Date
CHEVRON, U.S.A., INC.	8523-DB017	BELUGA RIVER FIELD	9/23/85	1/01/1990
CHEVRON, U.S.A., INC.	8523-DB003	SWANSON RIVER OIL FIELD	2/06/86	1/01/1990
CHEVRON, U.S.A., INC.	8323-DB035	BELUGA GAS FIELD	10/25/83	10/30/1988
CHEVRON, U.S.A., INC.	8523-DB032	SWANSON RIVER OIL FIELD	12/23/85	1/01/1990
CHEVRON, U.S.A., INC.	8523-DB031	SWANSON RIVER OIL FIELD	12/24/85	1/01/1990

APPENDIX DDD

CHEVRON, U.S.A., INC.	8523-DB029	SWANSON RIVER OIL FIELD	12/23/85	1/01/1990
CHEVRON, U.S.A., INC.	8523-DB028	SWANSON RIVER OIL FIELD	2/06/86	1/01/1990
MARATHON OIL CO.	8223-DB105	BEAVER CREEK OIL FIELD	11/08/82	10/01/1987
PHILLIPS PETROLEUM CO.	8523-DB022	KENAI PENINSULA BOROUGH	4/01/86	1/01/1990
UNION OIL CO. OF CALIF.	8423-DB015	KENAI GAS FIELD	6/07/85	1/01/1990

***-Out of over 2000 wells drilled on shore after the July 1973 statues of regulations went into effect only four disposal permits were applied for prior to Dave Brown going public with his complaints against the oil industry.**